75 **Topics in Current Chemistry**
Fortschritte der Chemischen Forschung

# Organic Chemistry and Theory



Springer-Verlag
Berlin Heidelberg GmbH 1978

This series presents critical reviews of the present position and future trends in modern chemical research. It is addressed to all research and industrial chemists who wish to keep abreast of advances in their subject.

As a rule, contributions are specially commissioned. The editors and publishers will, however, always be pleased to receive suggestions and supplementary information. Papers are accepted for "Topics in Current Chemistry" in English.

ISBN 978-3-662-15826-5 ISBN 978-3-540-35821-3 (eBook)
DOI 10.1007/978-3-540-35821-3

Library of Congress Cataloging in Publication Data. Main entry under title: Organic chemistry and theory. (Topics in current chemistry ; v. 75) Bibliography: p. Includes index. 1. Chemistry, Physical organic – Adresses, essays, lectures. I. Series. QD1.F58 vol. 75 [QD476] 540'.8s [547] 78-17664


Originally published by Springer-Verlag Berlin Heidelberg New York in 1978
Softcover reprint of the hardcover 1st edition 1978



2152/3140–543210

# Contents

Springer-Verlag — Postfach 105 280 · D-6900 Heidelberg 1
Telephone (0 62 21) 4 87-1 · Telex 04-61 723

Heidelberger Platz 3 · D-1000 Berlin 33
Telephone (0 30) 82 20 01 · Telex 01-83319

Springer-Verlag New York Inc. — 175, Fifth Avenue New York, NY 10010
Telephone 4 77-82 00

# Bimolecular Electron Transfer Reactions of the Excited States of Transition Metal Complexes

**Vincenzo Balzani, Fabrizio Bolletta, Maria Teresa Gandolfi, and Mauro Maestri**

Istituto Chimico "G. Ciamician" dell'Universitá, Bologna, Italy

**Table of Contents**

## List of Symbols

| | |
|---|---|
| bpy | = 2,2′-bipyridine |
| DMA | = N,N-dimethylaniline |
| DMF | = Dimethylformamide |
| DMSO | = Dimethylsulfoxide |
| EDTA | = Ethylendiaminetetraacetate ion |
| Et | = Ethyl |
| Me | = Methyl |
| NHE | = Normal Hydrogen Electrode |
| $P^{2+}$ | = 1,1′-dimethyl-4,4′-bipyridine$^{2+}$ (paraquat) |
| phen | = 1,10-phenanthroline |
| SCE | = Saturated Calomel Electrode |
| terpy | = 2,2′,2″-terpyridine |
| TPTZ | = 2,4,6-tripyridil-*s*-triazine |
| $\phi$ | = phenyl |

R1 =

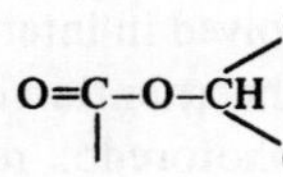

R2 =

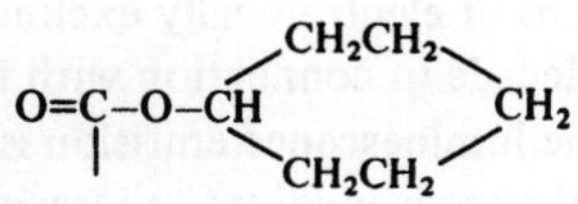

R3 =

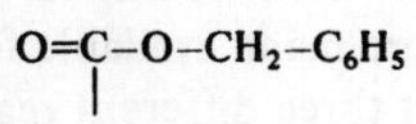

R4 =

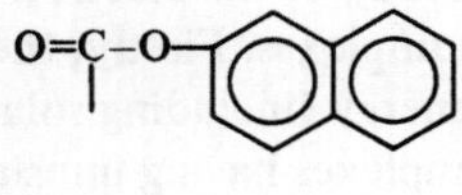

R5 =

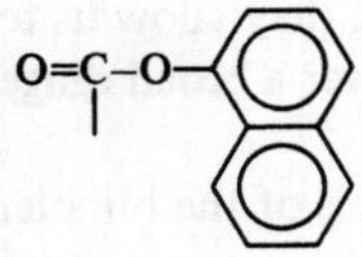

R6 =

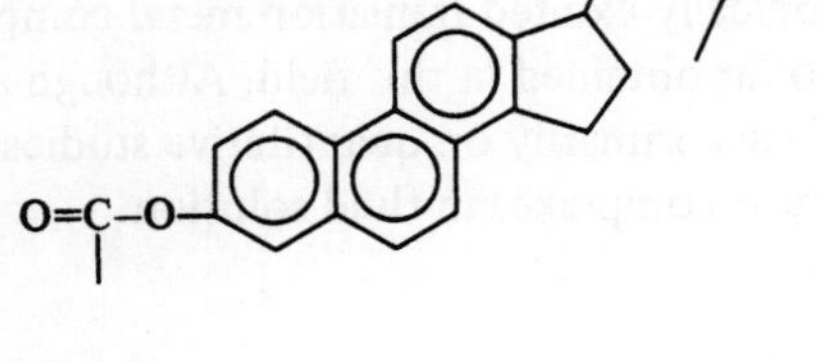

# 1. Introduction

It is well known that for each molecule there is the so-called ground state and there are many electronically excited states that can be obtained by visible or ultraviolet irradiation. In general, the ground state is the one that is responsible for the ordinary chemistry, while the electronically excited states are those responsible for the photochemical reactions. Each electronically excited state is virtually a new molecule with respect to the corresponding ground state and thus it can exhibit different chemical properties. This means that photochemistry offers a new dimension to chemistry[1], the excited state dimension, which, although mostly unexplored, is very promising for the progress of chemical research.

As will be shown later, one of the most important consequences of electronic excitation is that of increasing the electronic affinity and decreasing the ionization potential of a molecule. As a consequence, those electronically excited states which live long enough to encounter other species can easily be involved in intermolecular electron transfer reactions. Until a few years ago, however, the attention of inorganic photochemists was mainly focused on intramolecular photoredox reactions and ligand photosubstitution reactions[2, 3]. In the field of organic photochemistry, on the other hand, the electron transfer reactions of electronically excited states have been extensively investigated in the past decade in connection with fluorescence quenching and exciplex formation. As the luminescence emission is the most useful "handle" to establish whether an excited state is involved in intermolecular processes, the study of the electron-transfer reactions in transition metal photochemistry only began when luminescent complexes were made available for sensitization and quenching experiments[4]. There are at least three different reasons for the tremendous surge of interest and activity in the study of the electron-transfer reactions of electronically excited transition metal complexes. Firstly, these reactions are very promising for the conversion of light energy (including solar energy) into chemical energy. Secondly, they can lead to complexes having unusual oxidation states and thus unusual chemical properties. Finally, they allow us to check the theories of outer-sphere electron-transfer reactions over a broad range of free energy change.

In this review article, we discuss the fundamental basis of the bimolecular electron-transfer reactions of electronically excited transition metal complexes and then collect and examine the data so far obtained in this field. Although a wide range of systems are discussed, we focus primarily on quantitative studies, the majority of which involves Werner-type complexes in fluid solution.

# 2. Excited States of Transition Metal Complexes

The assignment of the various bands which appear in the absorption spectra of transition metal complexes is a very difficult problem because the absorption spectra reflect, of course, the complexity of the electronic structure of these molecules. From a photochemical point of view[2], it is convenient to make reference to schematic molecular orbital (MO) diagrams such as that shown in Fig. 1. In this very simplified diagram, each MO is labelled as metal (M) or ligand (L) according to its predominant localization. Thus, for example, the $a_{1g}$, $e_g$ and $t_{1u}$ bonding MO's, which result from the combination of metal and ligand orbitals of appropriate

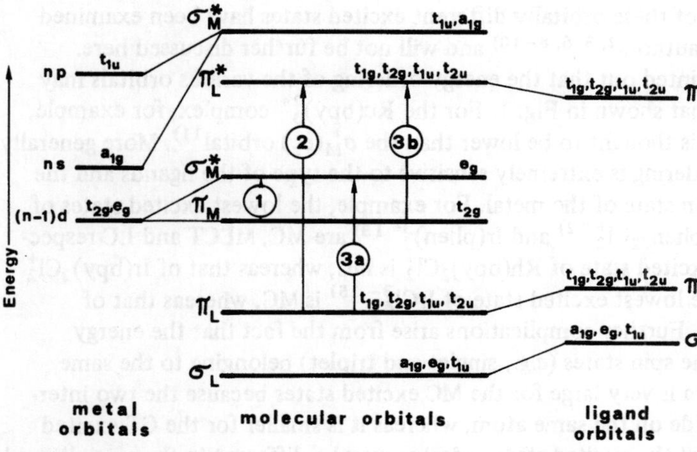


Fig. 1. Schematic orbital energy diagram representing various types of electronic transitions in octahedral complexes. A line connects an atomic orbital to that molecular orbital in which it has the greatest participation. 1: metal centered (MC) transitions; 2: ligand centered (LC) transitions; 3a: ligand-to-metal charge transfer (LMCT) transitions; 3b: metal-to-ligand charge transfer (MLCT) transitions

symmetry, are labelled with L because they receive the greatest contribution from the ligand orbitals. In the ground electronic configuration of transition metal complexes in their usual oxidation states, the $\sigma_L$ and $\pi_L$ orbitals are completely filled, the $\pi_M$ orbitals are either partially or completely filled and the higher orbitals are usually empty. Using the diagram of Fig. 1, it is possible to make a classification of the various electronic transitions according to the localization of the MO's involved. Specifically, we may identify three fundamental types of electronic transitions:

1. Transitions between MO's predominantly localized on the central metal; these are usually called *metal centered* (MC), ligand field or d–d transitions.

2. Transitions between MO's predominantly localized on the ligands. These are usually called *ligand centered* (LC) or intra-ligand transitions.

3. Transitions between MO's of different localization which cause the displacement of the electronic charge from the ligands to the metal or viceversa. These transitions are generally called *charge transfer* (CT) transitions and more specifically can be distinguished into *ligand-to-metal charge transfer* (LMCT) and *metal-to-ligand charge transfer* (MLCT) transitions.

Another important type of electronic transition is the *charge transfer to solvent* (CTTS) transition. Recently, it has also been found[5)] that electronic transitions may occur between MO's which are localized on different ligands of the same complex.

These transitions have been called *inter-ligand transmetallic charge transfer* transitions. Other types of transitions are present in polynuclear transition metal complexes[6, 7] but we will not deal with such compounds.

It must be pointed out that the classification into MC, LC and CT transitions (or excited states) is somewhat arbitrary and loses its meaning whenever the states involved cannot be described with localized MO configurations. The chemical and physical properties of these orbitally different excited states have been examined in detail by several authors[2, 3, 6, 8–10] and will not be further discussed here.

It should be pointed out that the energy ordering of the various orbitals may be different from that shown in Fig. 1. For the $Ru(bpy)_3^{2+}$ complex, for example, the $\pi_L^*(t_{1u})$ orbital is thought to be lower than the $\sigma_M^*(e_g)$ orbital[11]. More generally, the excited state ordering is extremely sensitive to the type of the ligands and the nature and oxidation state of the metal. For example, the lowest excited states of $Ir(phen)Cl_4^-$ [12], $Ir(phen)_2Cl_2^+$ [12] and $Ir(phen)_3^{3+}$ [13] are MC, MLCT and LC respectively; the lowest excited state of $Rh(bpy)_2Cl_2^+$ is MC, whereas that of $Ir(bpy)_2Cl_2^+$ is MLCT[14]; and the lowest excited state of $IrCl_6^{3-}$ [15] is MC, whereas that of $IrCl_6^{2-}$ [16] is LMCT. Further complications arise from the fact that the energy splitting between the spin states (*e.g.*, singlet and triplet) belonging to the same orbital configuration is very large for the MC excited states because the two interacting electrons reside on the same atom, whereas it is smaller for the CT excited states. It follows that the excited state ordering may be different in the spin-allowed and spin-forbidden manifolds. Finally, it should be recalled that the presence of heavy metals brings about a considerable degree of spin-orbit coupling and that this effect, being related to the central metal atom, is different for the different kinds of excited states. The LC excited states are scarcely affected, whereas for the MC and CT excited states it may become meaningless to talk about discrete spin states[17, 18]. In the current literature, however, the spin label is generally used even when its meaning is doubtful.

## 3. Quenching of Excited States

In a fluid solution, an excited state which has a lifetime long enough to encounter other species can be deactivated ("quenched") in a bimolecular process. The intimate mechanism of the quenching process is often difficult to elucidate[4, 19–21], but the final result (Fig. 2) is either (i) the electronic energy transfer from the excited state to the quencher, (ii) a chemical reaction between the excited state and the quencher or (iii) the deactivation of the excited state by some catalytic action of the quencher. Different quenching processes may also occur simultaneously, which causes noticeable complications in the study of these systems. The kinetic aspects of the interaction between an excited state and a quencher in a fluid solution have been extensively discussed[4, 20, 22–24]. The quenching processes are usually studied by measuring at different quencher concentrations one or more of the following quantities: (a) the emission or reaction efficiency of the excited state under continuous illumination or (b) the decay of the excited state emission or

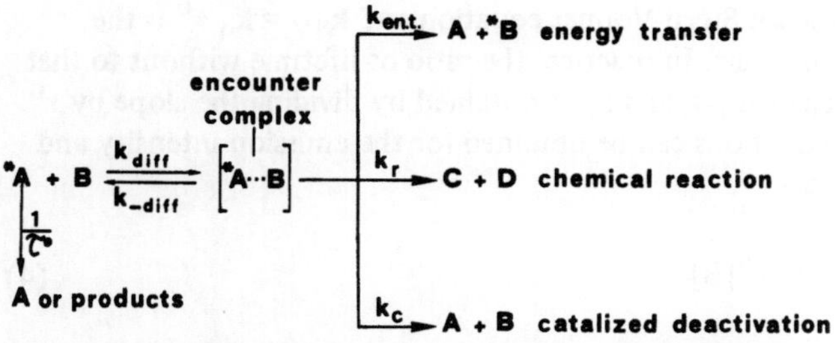


Fig. 2. Bimolecular quenching processes for an excited-state molecule

absorption after pulse excitation. The results obtained are usually discussed on the basis of a postulated mechanism which consists of a number of elementary transformations (steps)[25], each of which is characterized by a rate constant. In the relation between the experimental and mechanistic quantities, the experimental data generally appear as ratios of the value obtained in the absence of the quencher to the value obtained in the presence of a given amount of quencher. Consider for example the very simple scheme of Fig. 3, where $k_e$, $k_p$ and $k_d$ are the first order

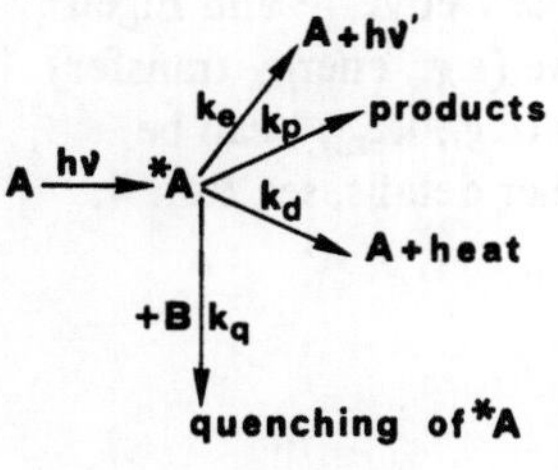


Fig. 3. Competition between intramolecular and intermolecular deactivations of an excited molecule

rate constants for luminescence, intramolecular reaction and radiationless deactivation of the excited state, and $k_q$ is the second order rate constant of the quenching reaction. In the absence of the quencher B the lifetime of the excited state is given by

$$\tau^0 = \frac{1}{k_e + k_p + k_d} \tag{1}$$

When the experiment is repeated in the presence of a given amount of quencher, the lifetime is shorter and is given by:

$$\tau = \frac{1}{k_e + k_p + k_d + k_q[B]} \tag{2}$$

From Eqs. (1) and (2) it follows that

$$\frac{\tau^0}{\tau} = 1 + k_q\,\tau^0\,[B] \tag{3}$$

Equation (3) is the well-known Stern-Volmer equation and $k_{SV} = k_q \tau^0$ is the Stern-Volmer quenching constant. In practice, the ratio of lifetime without to that with quencher is plotted against [B] and $k_q$ is obtained by dividing the slope by $\tau^0$. Analogous Stern-Volmer equations can be obtained for the emission intensity and the reaction quantum yield

$$I^0_{em}/I_{em} = \Phi^0_{em}/\Phi_{em} = 1 + k_q\tau^0[B] \tag{4}$$

$$\Phi^0_p/\Phi_p = 1 + k_q \tau^0 [B] \tag{5}$$

The experimental quenching constant $k_q$ is related to the rate constants of the quenching process in the encounter (Fig. 2) by Eq. (6);

$$k_q = k_{diff} \frac{(k_{en.t.} + k_r + k_c)}{k_{-diff} + (k_{en.t.} + k_r + k_c)} = k_{diff}\,\alpha \tag{6}$$

where $\alpha$ represents the efficiency of the quenching process during the encounter. The theoretical values for the diffusion rate constant, $k_{diff}$, and the encounter dissociation constant, $k_{-diff}$, can be calculated by means of the Debye[26] and Eigen[27] equations. When only one quenching mechanism is effective (*e.g.*, energy transfer) the rate constant in the encounter for that specific process (*e.g.*, $k_{en.t.}$) can be obtained. For limitations on the validity of Eq. (6) and other details, see Ref. 4.

## 4. Quenching by Chemical Reaction

The quenching of an excited state of a transition metal complex by chemical reaction can occur, in principle, by means of any of the intermolecular reactions which transition metal complexes are able to undergo. It should be noted, however, that intermolecular excited state reactions can only occur if they are fast enough to compete with the intramolecular deactivation modes of the excited state and with the other quenching processes (Fig. 2).

The most important classes of bimolecular reactions of transition metal complexes are ligand substitutions, reactions of the coordinated ligands and inner and outer sphere oxidation-reduction reactions[28].

Ligand substitution reactions are unlikely to be able to compete with the excited state decay unless the entering ligand is the solvent or a counter ion. In such a case, however, it is difficult to distinguish the bimolecular process from a unimolecular dissociative photoreaction[2]. Evidence in favor of a bimolecular ligand substitution process has recently been obtained for the following systems: $Mn(CO)_4NO\text{-}P\phi_3$[29], $Cr(bpy)_3^{3+} - H_2O$[30], $t\text{-}Cr(en)_2(NCS)_2^+ - OH^-$[31], $Cr(DMSO)_6^{3+} - N_3^-$[32] and $Cr(DMSO)_6^{3+} - SCN^-$[32]. More generally, photosolvation of Cr(III) complexes is thought to proceed *via* an associative-type mechanism[33] but definite proof is still lacking.

Some intermolecular reactions involving the coordinated ligands (*e.g.*, hydrogen or proton transfer) may be fast enough to compete with the excited state decay. However, except for a few cases [*e.g.*, $Ru(bpy)_2(CN)_2$ [34] and $Ru(bpy)_2(bpy\text{-}4,4'\text{-}(COOH)_2)^{2+}$ [35]] reactions of this kind have not yet been well documented for transition metal complexes, although they are very common for organic molecules[19, 36, 37].

Inner sphere oxidation-reduction reactions, which cannot be faster than ligand substitution reactions, are also unlikely to occur within the excited state lifetime. On the contrary, outer-sphere electron-transfer reactions, which only involve the transfer of one electron without any bond making or bond breaking processes, can be very fast (even diffusion controlled) and can certainly occur within the excited state lifetime of many transition metal complexes. In agreement with these expectations, no example of inner-sphere excited state electron-transfer reaction has yet been reported, whereas a great number of outer-sphere excited-state electron-transfer reactions have been shown to occur, as we well see later.

In dealing with neutral organic molecules in a fluid solution, it has been found that the formation of the electron transfer products in an excited state quenching process may be the result of a very complicated series of events. In particular, the encounter complex may give rise to an excited state complex (exciplex), *i.e.*, a situation in which the excited state and the quencher are linked by some kind of interaction[4, 38–44] (Fig. 4). Depending on its own properties and on the ex-

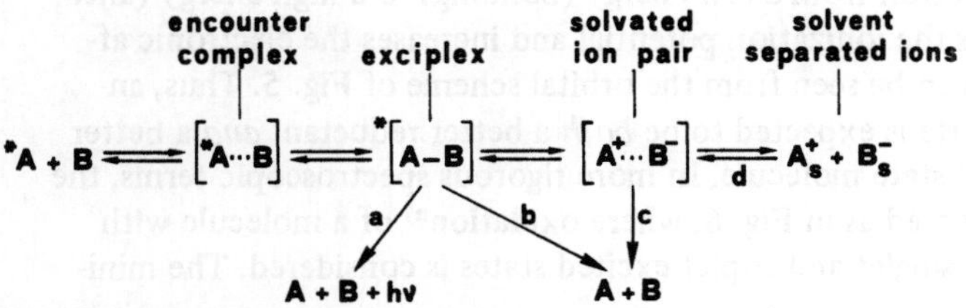


Fig. 4. Electron transfer quenching *via* exciplex formation. a: exciplex emission; b: exciplex radiationless deactivation; c: geminate ion recombination. For sake of simplicity, processes such as exciplex dissociation into A + *B, exciplex reactions other than the electron transfer one and direct formation of a solvated ion pair from the encounter complex have been omitted

perimental conditions, the exciplex can deactivate by luminescence, radiationless transitions or chemical reaction. When the chemical reaction is electron transfer, a solvated ion pair is formed at first, which then dissociates into solvent-separated ions. Exciplex emission may usually be observed in non-polar solvents but it disappears as the polarity of the solvent increases. Conversely, the solvent-separated ions can only be obtained in polar solvents. It is now generally accepted that the exciplex is an intermediate in many bimolecular photochemical transformations of organic molecules, although in most cases only indirect evidence is available. As far as the transition metal complexes are concerned, much less evidence is available for exciplex formation. The emission which was originally attributed[45] to an $*[Ir(phen)_2Cl_2^+$-naphthalene] exciplex has later been shown to be due to some kind of impurity[46]. Thus, the only documented cases seem to be those of excited metalloporphyrins with nitroaromatics, where non-emitting exciplexes have been detected by flash spectroscopy[41, 47, 48]. In general, it should be noted that both the structure of transition metal complexes (which in most cases are "spherical-type" molecules) and the nature of the solvents used (which have to be polar because the complexes are usually charged species) are not favorable to the formation of exciplexes.

## 5. Thermodynamic Aspects of Excited State Electron Transfer Reactions

The treatment of excited species as new chemical entities, and hence, the use of thermodynamics in dealing with their redox reactions is justified because in the condensed phase, the electronic relaxation times are usually several orders of magnitude longer than the time for thermal equilibration in all other degrees of freedom of a polyatomic molecule[49, 50].

The thermodynamics of excited states has been thoroughly discussed by Grabowski[50, 51] with particular reference to acid-base equilibria (Förster's cycle[52]). We will only consider here the problem of evaluating the difference between the redox potentials of $M^+/M$ and $M^+/{}^*M$ (or $M/M^-$ and ${}^*M/M^-$) couples where M is the ground state molecule and *M is an electronically excited state.

Excitation of an electron from a low energy (bonding) to a high-energy (antibonding) orbital reduces the ionization potential and increases the electronic affinity of a molecule, as can be seen from the orbital scheme of Fig. 5. Thus, an electronically excited state is expected to be *both* a better reductant *and* a better oxidant than the ground state molecule. In more rigorous spectroscopic terms, the situation may be represented as in Fig. 6, where oxidation[a] of a molecule with singlet ground state and singlet and triplet excited states is considered. The mini-

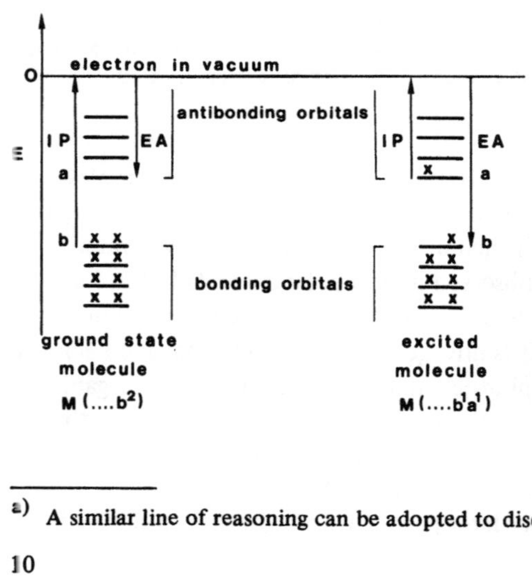


Fig. 5. Schematic orbital diagram showing the decrease of the ionization potential and the increase of the electronic affinity upon excitation of a molecule; b and a are the highest occupied (bonding) and the lowest unoccupied (antibonding) molecular orbitals in the ground state molecule

a) A similar line of reasoning can be adopted to discuss excited state reduction.

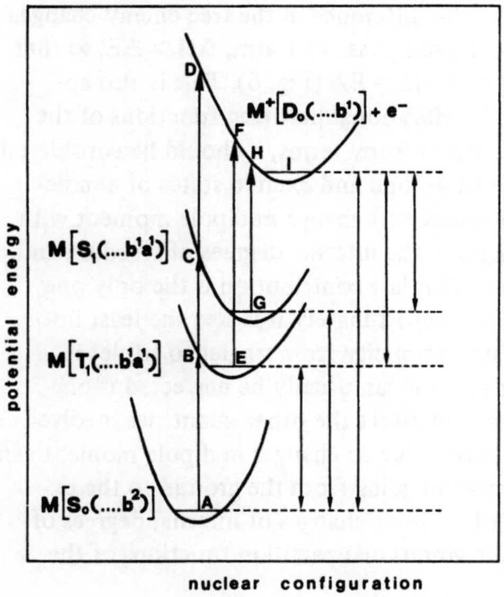


Fig. 6. Schematic representation of the excitation and ionization processes of a molecule. S, T and D indicate singlet, triplet and doublet states; for details, see text

mum of each curve represents the equilibrium configuration of the corresponding species including the solvation sphere. In this Figure, BA and CA represent the vertical (Franck-Condon) excitation energies from $S_0$ to $T_1$ and $S_1$, DA, FE and HG the vertical ionization energies of $S_0$, $T_1$ and $S_1$, EA and GA the zero-zero excitation energies of $T_1$ and $S_1$, and IA, IE and IG the equilibrium ionization potentials of $S_0$, $T_1$ and $S_1$. The electrochemical redox potentials of the couples $M(S_0)/M^+$, $M(T_1)/M^+$ and $M(S_1)/M^+$ *vs.* a reference couple $Q/Q^-$ are related to the free energy changes

$$\Delta G = \Delta H - T\,\Delta S \tag{7}$$

of the reactions

$$M(S_0) + Q \rightleftharpoons M^+ + Q^- \tag{8}$$

$$M(T_1) + Q \rightleftharpoons M^+ + Q^- \tag{9}$$

$$M(S_1) + Q \rightleftharpoons M^+ + Q^- \tag{10}$$

The difference in the free energy changes of reactions (8) and (9) can be expressed as follows

$$\Delta G[M(S_0), M^+] - \Delta G[M(T_1), M^+] = \Delta H[M(S_0), M^+] - \Delta H[M(T_1), M^+] - \\ - T\{\Delta S[M(S_0), M^+] - \Delta S[M(T_1), M^+]\} \tag{11}$$

An analogous equation can be obtained for the difference in the free energy changes between reactions (8) and (10). In the condensed phase at 1 atm, $\Delta H \simeq \Delta E$, so that at 0 K $\Delta H[M(S_0), M^+] - \Delta H[M(T_1), M^+] = IA - IE = EA$ (Fig. 6). This is also approximately true at room temperature if the vibrational partition functions of the two states are not very different[50]. As to the entropy terms, it should be considered that the difference in the entropy content of ground and excited states of a molecule may be due to three different contributions: (i) change in dipole moment with consequent changes in solvation, (ii) changes in the internal degrees of freedom and (iii) changes in orbital and spin degeneracy. This last contribution is the only one that can be straightforwardly calculated, but unfortunately it is also the least important one in most cases. For a change in multiplicity from singlet to triplet it amounts to 0.03 eV at 25 C, which means that it can usually be neglected if one considers the experimental uncertainties which affect the other quantities involved in these calculations. The entropy contribution due to changes in dipole moment can be calculated if the change in dipole moment in going from the ground to the excited state is known[50]. Finally, the contribution of changes of internal degrees of freedom is difficult to evaluate because the vibrational partition functions of the excited states are usually unknown.

Qualitatively speaking, the various types of excited states of the transition metal complexes are expected to behave differently as far as the entropy contributions (i) and (ii) are concerned. The metal centered excited states are expected to exhibit substantial changes in the metal-ligand bond frequencies[b)] but small or no changes in the dipole moment with respect to the ground state. The ligand-to-metal charge transfer excited states are expected to exhibit small or large changes in the metal-ligand frequencies depending on whether the electron is promoted to $\pi$ or $\sigma^*$ metal orbitals; the intra-ligand frequencies, of course, may also change in a way depending on the ligand nature. The change in the dipole moment upon LMCT transitions may be zero or large, depending on whether the promoted electron comes from an orbital which is delocalized on all the ligands or which is localized on one particular ligand. In the metal-to-ligand charge transfer excited states promotion of an electron to a $\pi$ antibonding orbital of the ligand(s) is not expected to affect greatly the metal-ligand bond frequencies but it may have important consequences on the ligand vibrations; the change on dipole moment can be zero or large as in the previous case, depending on the extent of localization of the promoted electron. Finally, the ligand centered excited states will exhibit small or no changes in dipole moment, whereas the frequency changes in the ligand vibrations will depend on the specific nature of the ligand.

It is well known that the change in shape, size and solvation of an excited state with respect to the ground state causes a shift (Stokes shift) between absorption and emission[2, 3, 9, 10]. When the Stokes shift is small the changes in shape, size

---

b) This is true for MC excited states obtained from $\pi_M \rightarrow \sigma^*_M$ electronic transitions such as the $\pi_M(t_{2g}) \rightarrow \sigma^*_M(e_g)$ one in octahedral $d^6$ complexes[2, 3]. Intraconfiguration MC transitions such as that leading from the $^4A(t^3_{2g})$ ground state to the $^2E(t^3_{2g})$ excited state of Cr(III) complexes do not involve electron promotion to antibonding orbitals and thus they are not expected to cause appreciable changes in the metal-ligand frequencies[2, 3].

and solvation are also small and the difference between the entropy content of the excited state and ground state may probably be neglected. In such a case, the redox potentials of the excited state are fairly well approximated by the following equations (see Fig. 7):

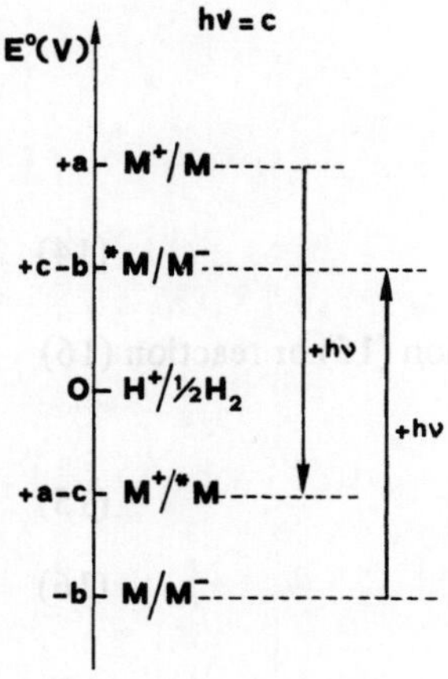


Fig. 7. Schematic diagram showing the difference in the redox properties of the ground and excited molecule according to Eqs. (12) and (13); c is the one-electron potential corresponding to the zero-zero spectroscopic energy of the excited state

$$E^0\,(M^+/{*M}) = E^0\,(M^+/M) - E_{0-0}(M-{*M}) \qquad (12)$$

$$E^0\,({*M}/M^-) = E^0\,(M/M^-) + E_{0-0}(M-{*M}) \qquad (13)$$

where $E_{0-0}(M-{*M})$ is the one-electron potential corresponding to the zero-zero spectroscopic energy of the excited state. More generally, as the entropy corrections are practically impossible to evaluate, Eqs. (12) and (13) are commonly used to obtain an estimate of excited state redox potentials.

Excited state potentials may also be evaluated from quenching measurements using a series of quenchers of graded potentials. This approach has been used for evaluating the $*Ru(bpy)_3^{2+}$ oxidation potential[53] as well as the $*Cr(bpy)_3^{3+}$, $*Ru(bpy)_3^{2+}$ and $*Ir(Me_2phen)_2Cl_2^+$ reduction potentials[54], as we will see in detail later. A more direct approach to excited state potentials is that of electrochemical measurements on irradiated solutions using suitable semiconductor electrodes[55, 56]. This last technique, which has only been used in a few cases[57, 58] because of intrinsic difficulties, offers the advantage that it does not require a luminescent excited state.

In organic molecules there is generally a great separation between oxidation states, so that an excited organic molecule can usually serve as either an electron donor or an electron acceptor but not both. In transition metal complexes the presence of redox sites on both metal and ligand offers additional possibilities not available to either simple metal ions or organic molecules. The oxidation states are often closely spaced, so that the case is very common in which an excited state can be used as both an electron donor and an electron acceptor. As can be seen from Fig. 7, when the oxidation and reduction potentials of a molecule are close enough, it is common to have light-driven crossing of the excited state potentials. In other words, the reduction potential of the $^*M/M^-$ couple may become more positive than the reduction potential of the $M^+/^*M$ couple. When this happens (Fig. 7) the stability of the ground state towards disproportionation is lost in the excited state:

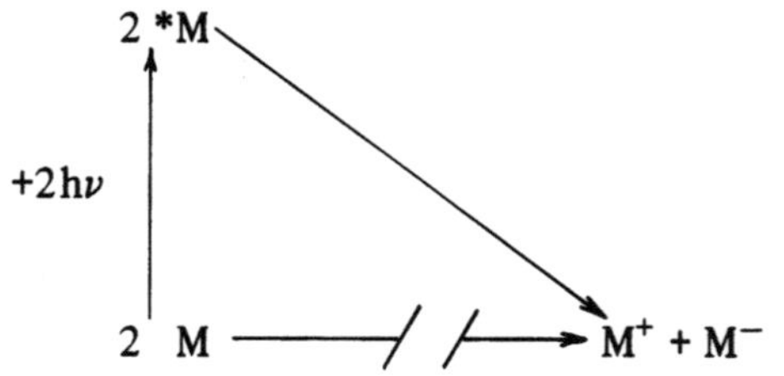


(14)

Depending on the values of a, b and c in Fig. 7, either reaction (15) or reaction (16) will be thermodynamically favored. Reaction (15)

$$^*M + M \longrightarrow M^+ + M^- \qquad (15)$$

$$M^+ + M^- \longrightarrow {}^*M + M \qquad (16)$$

represents a self-quenching *via* electron transfer, whereas reaction (16) is a conproportionation reaction with formation of an excited state. Reaction (16) has been observed for both organic molecules[59, 60] and transition metal complexes[11]. The best way to obtain this reaction is to produce the oxidized and reduced forms by alternating current electrolysis of the species M ("electrogenerated chemiluminescence").

## 6. Kinetic and Theoretical Aspects of Outer-Sphere Electron Transfer Reactions

### 6.1. Thermal Reactions

The oxidation-reduction reactions of transition metal complexes have been the object of extensive investigations in the last twenty years. Since comprehensive reviews are available[61–68] on the experimental and theoretical aspects of these reactions, we will only deal with those aspects which are more strictly related to our topic.

It has been shown that the oxidation-reduction reactions of transition metal complexes can occur through two different mechanisms: (i) inner-sphere mechanism, when the two reactants share one or more ligands of their first coordination spheres in the activated complex and (ii) outer-sphere mechanism, when the first coordination spheres of the two reactants are left intact (as far as the number and kind of ligands present are concerned) in the activated complex. As mentioned before, the inner-sphere reactions cannot be faster than ligand substitution and thus they are unlikely to occur within the excited state lifetime. The outer-sphere reactions, on the contrary, can be very fast and for many transition metal complexes they can certainly compete with excited state decay.

The simplest outer-sphere electron transfer reaction is the so-called *self-exchange reaction,* where the two reactants are converted one into the other by electron transfer. For example:

$$Fe(CN)_6^{4-} + Fe(CN)_6^{3-} \rightleftharpoons Fe(CN)_6^{3-} + Fe(CN)_6^{4-} \quad (17)$$

The rate of self-exchange reactions can generally be measured by isotopic tracer methods, but in several cases other techniques (optical rotation, nmr, epr) are more useful. Reactions like (18),

$$Fe(CN)_6^{4-} + Co(phen)_3^{3+} \rightleftharpoons Fe(CN)_6^{3-} + Co(phen)_3^{2+} \quad (18)$$

where the products are different from the reactants, are usually called *cross-reactions.*

In an outer-sphere electron transfer reaction like that shown by Eq. (19) it is

$$A + B \rightleftharpoons A^+ + B^- \quad (19)$$

convenient to distinguish three elementary steps:

$$\underset{\text{reactants}}{A + B} \rightleftharpoons \underset{\substack{\text{precursor}\\ \text{complex}}}{A \cdots B} \rightleftharpoons \underset{\substack{\text{successor}\\ \text{complex}}}{A^+ \cdots B^-} \rightleftharpoons \underset{\text{products}}{A^+ + B^-} \quad (20)$$

The first step involves the formation of the precursor complex, where the reactants maintain their identity. In the second step there is, as we will see later, reorganization of the inner coordination shells as well as of the solvation spheres of the reactants so as to obtain a nuclear configuration appropriate to the activated complex through which the precursor complex is transformed into the successor complex. The electron transfer usually occurs during the latter stages of this reorganization process. The activated complex deactivates to form the successor complex if electron transfer has occurred or to reform the precursor complex if electron transfer has not occurred. The electron distribution in the successor complex corresponds to that of the products, so that the third step is simply the dissociation of the successor complex to form the separated products.

In the theoretical approach to chemical kinetics it is customary to represent chemical reactions in terms of the motion of a point representing the system on a potential energy surface. Usually, only the most important path (reaction coordinate) leading from the reactants to the products is considered, so as to have a two-dimension diagram. In principle, the potential energy surface may be constructed by solving the Schrödinger equation for the systems using the Born-Oppenheimer approximation. In practice, however, only approximate potential energy surfaces can be obtained by using approximate wave functions. For an outer-sphere electron transfer reaction the electronic wave function of the separated ground state reactants, $\psi_R$, can be used to obtain a zero-order energy surface, $E_R$, and the electronic wave function of the separated products, $\psi_P$, can similarly be used to obtain another zero-order wave function, $E_P$. Each one of these surfaces will have its own valleys, corresponding to the more stable nuclear configurations of the reactants and products, respectively. A profile of these surfaces for a self-exchange reaction is shown in Fig. 8. The Franck-Condon principle requires that the nuclear positions and nuclear velocities remain essentially unchanged during the electronic transition. The occurrence of the electron-transfer process when the reactants are in their equilibrium configuration needs a large input of energy during the time required for the electronic transition because the products can only be obtained in a very high energy state at the equilibrium configuration of the reactants. This is only possible in a photochemical process (Fig. 8)[c]. In a thermal process, adjustment of the

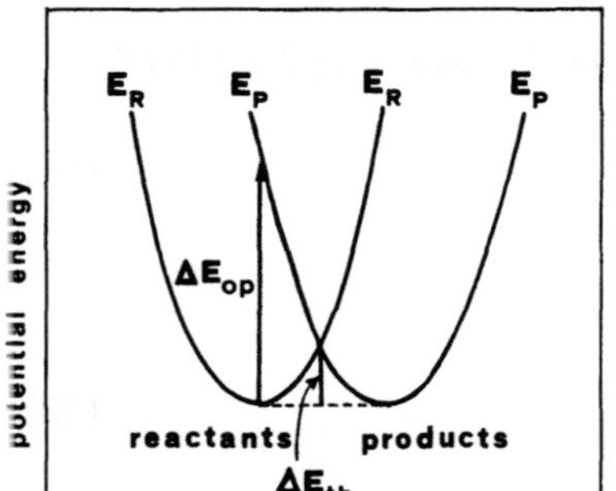


Fig. 8. Profile of the potential energy surface as a function of the nuclear configuration for a self-exchange reaction.
$\Delta E_{th}$ is the activation energy for the thermal electron transfer. $\Delta E_{op}$ is the Franck-Condon energy for the photochemical electron transfer

inner coordination shells of the two reactants and rearrangement of the outer solvation shells to some non-equilibrium configuration are required prior to electron transfer. The zero-order surfaces will intersect at those nuclear configurations for which $E_R = E_P$. In the intersection region, however, the two zero-order states, $\psi_R$ and $\psi_P$, will interact so that the system can no longer be described by the zero-order wave functions. A better, though still approximate, description is given by a linear combination of $\psi_R$ and $\psi_P$. Application of the variation method then gives the first-order energies $E_+$ and $E_-$ of the two new states, $\psi_+$ and $\psi_-$ (Fig. 9). If the overlap integral between $\psi_R$ and $\psi_P$ is $\simeq 0$, the separation energy between $E_+$ and $E_-$ in the intersection region is twice the interaction energy between $\psi_R$ and $\psi_P$, $H_{RP}$. The difference between the minimum potential energy of the reactants and the potential energy of the lowest pass through the intersection region is equal to the potential energy of activation for the reaction (Fig. 8). Theoretical approaches to

c) The optical transition corresponding to this process can be observed in binuclear mixed-valence ions[69, 70].

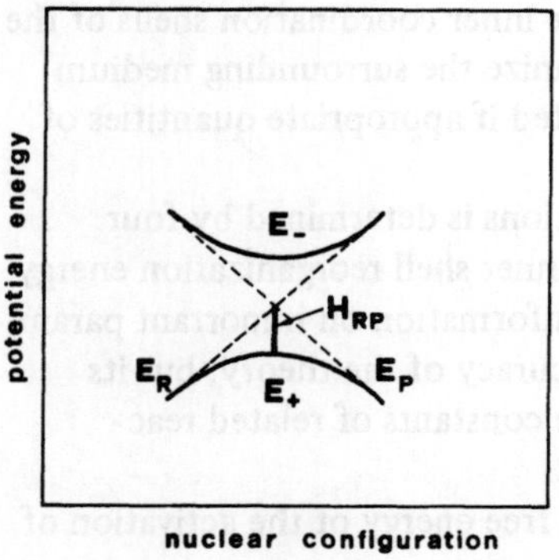


Fig. 9. Interaction between the zero-order potential energy surfaces in the intersection region. $E_+$ and $E_-$ are the first-order surfaces. $H_{RP}$ is the interaction energy

the behavior of the system in the intersection region[65, 68] show that when the interaction energy is greater than a few tenths of kcal/mol, the splitting of the first-order surfaces will be sufficiently large so that the point representing the system will remain on the lower first-order energy surface as it moves along the reaction coordinate and the probability of reactants being converted into products in the transition state is unity. In such a case, the reaction is called *adiabatic.* The interaction energy is very small when the separation of the reactants is large or when the electron transfer is spin or symmetry forbidden.

The most detailed theoretical treatment of outer-sphere electron-transfer reactions has been given by Marcus[71–75], who has derived the following expression for the rate constant for an oxidation-reduction reaction:

$$k = pZe^{-(w_r + \Delta G^{**})/RT} \tag{21}$$

where p is the probability of reactants being converted into products (equal to unity in adiabatic reactions), Z is the collision frequency between two uncharged reactants in a solution ($\sim 10^{11} M^{-1} s^{-1}$), $w_r$ is the work required to bring the reactants from an infinite distance apart to their separation in the activated complex and $\Delta G^{**}$ is the free energy required to reorganize the coordination shells of the reactants prior to the electron transfer. The work term $w_r$, which is related to the stability of the precursor complex, can be approximately calculated from equations of the kind given by Eq. (22)[76], where $z_1$ and $z_2$ are the formal charges of the reactants,

$$w_r = \frac{z_1 z_2 e^2}{D_s a} e^{-a\chi} \tag{22}$$

e is the electron charge, $D_s$ is the static dielectric constant of the medium, a is the distance between the centers of the reactants and the exponential is the Debye-Hückel term with $\chi$ as the reciprocal Debye radius. The free energy of reorganization, $\Delta G^{**}$, is actually given by

$$\Delta G^{**} = \Delta G_i^{**} + \Delta G_0^{**} \tag{23}$$

where $\Delta G_i^{**}$ is the energy required to reorganize the inner coordination shells of the reactants and $\Delta G_0^{**}$ is the energy required to reorganize the surrounding medium (outer shells). Both $\Delta G_i^{**}$ and $\Delta G_0^{**}$ can be calculated if appropriate quantities of the system are known[63, 68].

The rate of self exchange electron transfer reactions is determined by four factors: the interaction energy, the work term, the inner shell reorganization energy and the solvent reorganization energy. The lack of information on important parameters does not allow general tests of the absolute accuracy of the theory, but its formal validity can be proved by comparing the rate constants of related reactions[65, 68].

Marcus[72, 73] and Sutin[77] have shown that the free energy of the activation of a cross-reaction [Eq. (24)]

$$A_1 + A_2 \underset{}{\overset{k_{12}}{\rightleftharpoons}} A_1^+ + A_2^- \tag{24}$$

can be expressed by

$$\Delta G_{12}^{**} = \frac{(\Delta G_1^{**} + \Delta G_2^{**})}{2} + \frac{\Delta G_r^0}{2} + \frac{(\Delta G_r^0)^2}{8\,(\Delta G_1^{**} + \Delta G_2^{**})} \tag{25}$$

where $\Delta G_1^{**}$ and $\Delta G_2^{**}$ refer to the appropriate self-exchange reactions [Eqs. (26) and (27)] and $\Delta G_r^0$ is the free energy change

$$A_1 + A_1^+ \overset{k_1}{\rightleftharpoons} A_1^+ + A_1 \tag{26}$$

$$A_2 + A_2^- \overset{k_2}{\rightleftharpoons} A_2^- + A_2 \tag{27}$$

for the cross reaction when the two reactants are at a distance r apart. In deriving Eq. (25), the assumption is made that the splitting at the intersection is small enough so that it can be neglected in calculating the free energy of activation but large enough so that the value of p is close to unity. Equation (25) is often written as

$$\Delta G_{12}^{**} = \frac{\lambda_{12}}{4}\left(1 + \frac{\Delta G_r^0}{\lambda_{12}}\right)^2 \tag{28}$$

where

$$\lambda_{12} = 2\,(\Delta G_1^{**} + \Delta G_2^{**}) \tag{29}$$

is the so-called reorganizational parameter. For a self-exchange reaction, using Eqs. (23) and (29) one obtains:

$$\lambda = 4\Delta G^{**} = 4(\Delta G_i^{**} + \Delta G_0^{**}) = \lambda_i + \lambda_0 \tag{30}$$

It should be recalled that the measured free energy of activation, $\Delta G^*$, includes the work term, *i.e.*, the energy required to bring the reactants from an infinite distance apart to their separation distance in the activated complex:

$$\Delta G^{**} = \Delta G^* - w_r \tag{31}$$

Similarly, the standard free energy change for the reaction when the two reactants are infinitely far apart, $\Delta G^0$, includes the work terms relative to reactants and products:

$$\Delta G_r^0 = \Delta G^0 - w_r + w_p \tag{32}$$

Under particular conditions[65, 68, 79)], Eq. (25) can be related to the appropriate rate constants yielding the following expression:

$$k_{12} = (k_1 k_2 K_{12} f)^{1/2} \tag{33}$$

where $\log f = (\log K_{12})^2/4 \log (k_1 k_2/Z^2)$ with $K_{12}$ the equilibrium constant for the cross reaction. The rate constants for a number of cross reactions between transition metal complexes calculated by Eq. (33) have been found to agree well with the experimental values[65, 66, 68, 79)]. Recently, the Marcus relationships have also been applied to reactions between transition metal complexes and organic redox systems[80)].

It is interesting to examine Eq. (25) in more detail. When $\Delta G_r^0 \ll 8(\Delta G_1^{**} + \Delta G_2^{**})$, Eq. (25) reduces to

$$\Delta G_{12}^{**} = \frac{\Delta G_1^{**} + \Delta G_2^{**}}{2} + \frac{\Delta G_r^0}{2} = \frac{\lambda_{12}}{4} + \frac{\Delta G_r^0}{2} \tag{34}$$

Under such conditions a linear relationship with slope 0.5 is predicted (and often observed) between the free energies of activation and the standard free energy changes for a series of reactions for which $\lambda_{12}$ is a constant. $\Delta G_{12}^{**}$ will reach a minimum for $\Delta G_r^0 = -\lambda_{12}$, and then it should increase when $\Delta G_r^0$ becomes large (with respect to $\lambda_{12}$) and negative. This means that the rate constants of the electron transfer step for cross reactions between a single reductant $A_1$ and a homogeneous series of oxidants ($A_{2a}$, $A_{2b}$, etc.) should increase up to a maximum and then decrease with increasing oxidation potential of the oxidant (Fig. 10). The region in which the rate constant should decrease with increasing $|-\Delta G_r^0|$ is usually called the *Marcus inverted or abnormal region.* It should be noted that Eq. (25) applies to the free energy of activation and the free energy for the actual electron transfer step, which does not necessarily coincide with the free energy of activation of the overall reaction ($\log k_{12}$) and the overall free energy change ($\log K_{12}$). The increase in the activation free energy when the free energy change becomes large and negative is also apparent from Fig. 11, where for large and negative $\Delta G_r^0$ values the crossing between the surfaces of the reactants and products occurs in the left-hand side of the surface of the reactants at progressively higher energies. It is also apparent that

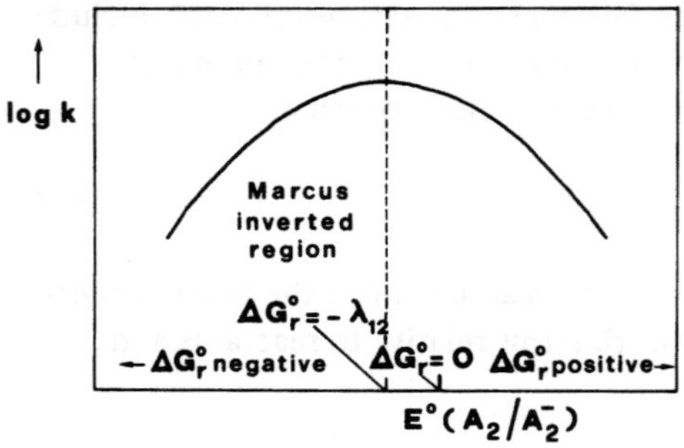


Fig. 10. Expected relationship, according to Eq. (28), between the rate constant of the electron transfer steps and the reduction potential of the oxidant for a cross reaction of the type of reaction (24). The reorganizational parameter $\lambda_{12}$ is given by Eq. (29)

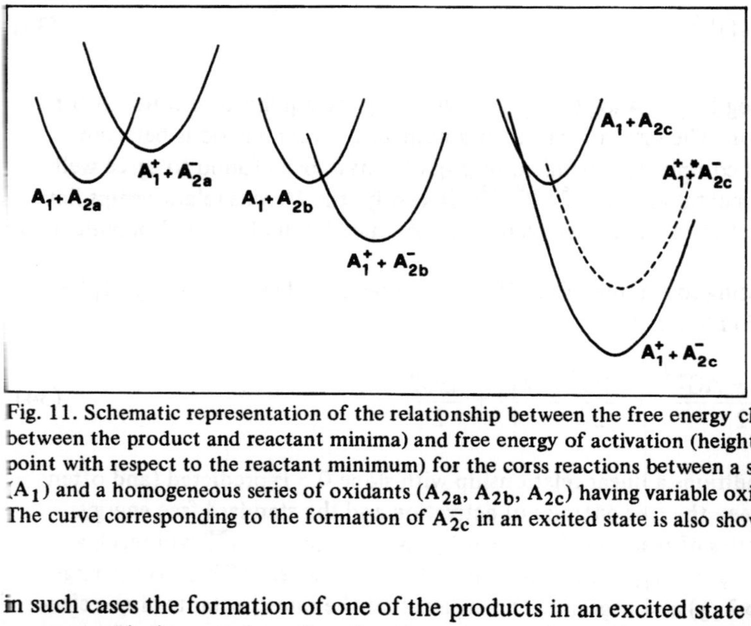


Fig. 11. Schematic representation of the relationship between the free energy change (difference between the product and reactant minima) and free energy of activation (height of the crossing point with respect to the reactant minimum) for the corss reactions between a single reductant ($A_1$) and a homogeneous series of oxidants ($A_{2a}$, $A_{2b}$, $A_{2c}$) having variable oxidation potential. The curve corresponding to the formation of $A_{2c}^-$ in an excited state is also shown

in such cases the formation of one of the products in an excited state may require a comparable (or even lower) activation energy and may thus compete with the formation of ground state products. Actually, formation of excited states in suitable electron transfer reactions is proven by the observation of luminescence emission (chemiluminescence)[81, 82]. A typical example is given by the reaction of $Ru(bpy)_3^{3+}$ with $e_{aq}^-$ [78, 83], which leads partly to the ground state and partly to excited state(s) of $Ru(bpy)_3^{2+}$ (Section 11.3).

Few opportunities for systematic exploration of the inverted region arise when both reactants are in their ground electronic state, since $\Delta G_r^0$ is not sufficiently negative. Using a very powerful reductant like $e_{aq}^-$, however, produced evidence of a decrease of the rate constant in the inverted region [see, for example, Ref. (84)]. Excited state systems with small reorganization energies provide excellent probes of this region. Rehm and Weller[85, 86] in their classic papers on the electron-transfer quenching of hydrocarbon fluorescence did not find evidence of a decrease of the rate constant in the inverted region. Ballardini *et al.* [54] studied the electron-transfer quenching of $*Cr(bpy)_3^{3+}$ and $*Ir(Me_2phen)_2Cl_2^+$ by aromatic amines and methoxybenzenes and again did not find evidence for a decrease of the rate constant in the inverted region (for details, see Section 9.6). Evidence for diminished rate constants in the inverted region has been found by Creutz and Sutin[87] in the electron-transfer quenching of excited polypiridine ruthenium(II) complexes by polypirydine complexes of Ru(III), Cr(III) and Os(III) (Section 9.6). Even in this case, however, the rate constants were many orders of magnitude above those predicted by Eq. (28). The disagreement between the experimental results and the prediction of the conventional Marcus theory in the inverted region has prompted several authors to refine the theory or to explore other types of approaches. It has been recognized that adiabatic approximation cannot be used in the inverted region and recent theoretical work[88–92] has suggested that in such a region the usual reorganizational barrier to electron transfer can be circumvented by nuclear tunneling.

## 6.2. Excited State Reactions

The participation of excited states as either reactants or products in electron-transfer reactions causes some additional problems with respect to the ground state reactions. As we have seen in Section 5, an excited state may differ from the corresponding ground state in size, shape and dipole moment. Thus, the reorganization parameter $\lambda$ will generally be different for ground and excited states of the same molecule. For example, if the solvation arrangement in the excited state is closer to the solvation arrangement of the product, $\lambda_0$ will be smaller for the excited state state than for the ground state. As the Stokes shift between ground state absorption and excited state emission is related to the differences in the ground state and excited state nuclear coordinates, it is also expected to be related in some way to the difference in the reorganizational parameters of ground and excited states. Such a relation, however, is difficult to find except in the trivial case of a Stokes shift equal to zero, *i.e.*, when the ground and excited states having the same shape, size and solvation, must also have the same reorganizational parameters. Let us consider, for example, the two self-exchange reactions

$$ML_n^{2+} + ML_n^{3+} \rightleftharpoons ML_n^{3+} + ML_n^{2+} \tag{35}$$

$$*ML_n^{2+} + ML_n^{3+} \rightleftharpoons ML_n^{3+} + *ML_n^{2+} \tag{36}$$

and try to evaluate the differences in the reorganizational parameters (Fig. 12). If we assume that: (i) the coordinate which is responsible for the Stokes shift coincides with the reaction coordinate, (ii) each species is described by the same harmonic function and (iii) the zero-point energy is negligible, the following relation can be obtained:

$$\Delta E_a = E_a - {}^*E_a = \frac{\Delta h\nu}{8}\left(2\,\frac{r_2 - r_1}{{}^*r_1 - r_1} - 1\right) = \frac{\Delta h\nu}{8}\left\{\pm\,\frac{4\sqrt{2}\sqrt{E_a}}{\sqrt{\Delta h\nu}} - 1\right\} \tag{37}$$

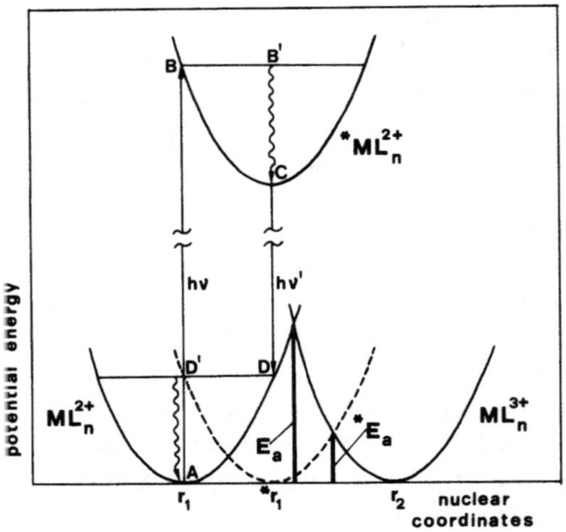


Fig. 12. Schematic representation of the relation between the spectroscopic Stokes shift and the difference in the reorganization barriers for ground and excited state electron transfer reactions. AB = absorption; CD = emission; B'C + D'A = $\Delta h\nu$ (Stokes shift); $2E_a$ = reorganization barrier for reaction (35); $2{}^*E_a$ = reorganization barrier for reaction (36)

where $2E_a$ and $2{}^*E_a$ are the reorganization barriers for reaction (35) and (36), $\Delta h\nu$ is the Stokes shift and $r_1$, ${}^*r_1$ and $r_2$ are the equilibrium coordinates of $ML_n^{2+}$, ${}^*ML_n^{2+}$ and $ML_n^{3+}$ (Fig. 12). The plus or minus term in Eq. (37) reflects the situations in which ${}^*r_1$ and $r_2$ are on the same or on different sides of $r_1$ along the reaction coordinate. Equation (37) is relatively simple but the assumption of the same harmonic function for the ground state, the excited state and the oxidized form is clearly unreal. The coincidence between the coordinate responsible for the Stokes shift and that along which electron transfer takes place may also be difficult to establish.

Another important difference between ground-state and excited state electron-transfer reactions is that the orbitals involved are different in the two cases, which may cause a different interaction energy. Detailed examples will be given in Section 9.6.

Rehm and Weller[85, 86] have discussed in detail the kinetic and thermodynamic aspects of the quenching of an excited state by electron transfer. The treatment below follows that given by these authors. Consider the kinetic scheme shown in Fig. 13, where electron transfer quenching of an excited state molecule, ${}^*A_1$, by a ground state molecule, $A_2$, leads to the redox products $A_1^+$ and $A_2^-$, which could also be obtained in the ground state electron-transfer process. In this scheme, $k_d$, $k_{-s}$ and $k'_d$ are the appropriate diffusion rate constants, $k_{-d}$, $k_s$ and $k'_{-d}$ are the dissociation rate constants of the precursor and successor complexes, $k_e$, $k_{-e}$, $k'_e$ and $k'_{-e}$ are the rate constants of forward and back electron transfer for the ground and excited state reactions, respectively, and $1/\tau^0$ is the rate constant of intramolecular deactivation of the excited state. It should be noted that, in dealing with transition metal complexes, $A_1$ and $A_2$ may be charged species so that $A_1^+ \cdots A_2^-$, in general, will not be a "one plus-one minus" ion pair, as it usually happens with

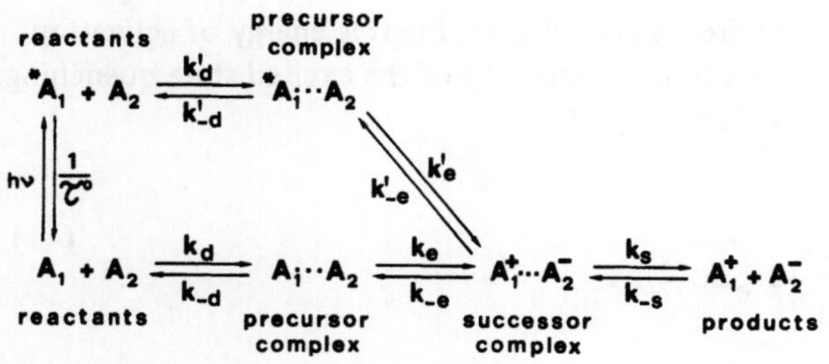


Fig. 13. Kinetic scheme for an excited state electron transfer reaction. For details, see text

organic molecules[85, 86]. Let us consider a case in which electron transfer in the forward direction is thermodynamically forbidden when the two molecules are in their ground states, whereas it is allowed when $A_1$ is excited (Fig. 14). Using the Stern-

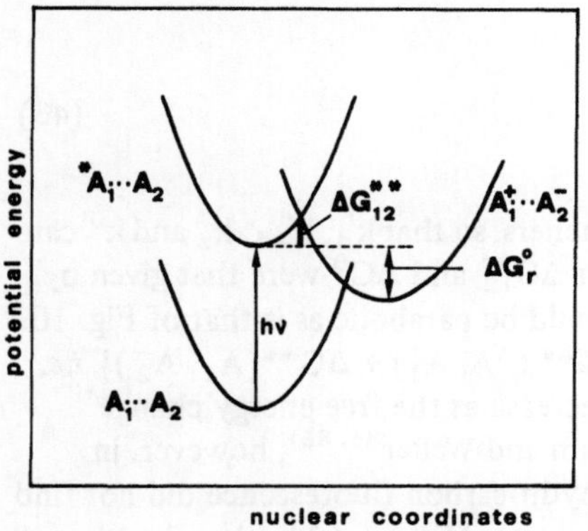


Fig. 14. Energy profiles for the electron transfer steps of the kinetic scheme shown in Fig. 13

Volmer relationship and steady-state approximations, the kinetic scheme of Fig. 13 leads to Eq. (38) where $k_q$ is the

$$k_q = \frac{k'_d}{1 + \frac{k'_{-d}}{k'_e}\left(1 + \frac{k'_{-e}}{k_x}\right)} \tag{38}$$

observed bimolecular quenching constant (Section 3) with $k_x \simeq k_{-e}$ when $k_{-e} \gg k_s$ or when $k_{-s}$ is the only path through which $A_1^+$ and $A_2^-$ can disappear, and $k_x \simeq k_s$ when $k_s \gg k_{-e}$ and some reaction other than $k_{-s}$ rapidly consumes $A_1^+$ and/or $A_2^-$. From Fig. 14 it is evident that

$$\frac{k'_{-e}}{k'_e} = \frac{1}{K'_{e,-e}} = e^{\Delta G_r^0/RT} \quad \text{and} \tag{39}$$

$$k'_e = k^0\, e^{-\Delta G_{12}^{**}/RT} \tag{40}$$

where $\Delta G_r^0$, $\Delta G_{12}^{**}$ and $k^0$ are the free energy change, the free energy of activation and the frequency factor of the electron transfer step of the excited state quenching. Thus, Eq. (38) can be transformed into Eq. (41):

$$k_q = \frac{k'_d}{1 + \frac{k'_{-d}}{k_x} e^{\Delta G_r^0/RT} + \frac{k'_{-d}}{k^0} e^{\Delta G_{12}^{**}/RT}} \tag{41}$$

Equation (41) reduces to Eq. (42) when $\Delta G_r^0$ becomes large and negative and to Eq. (43) when $\Delta G_r^0$ becomes large and positive:

$$k_q = \frac{k'_d}{1 + \frac{k'_{-d}}{k^0} e^{\Delta G_{12}^{**}/RT}} \tag{42}$$

and

$$k_q = \frac{k'_d k^0 k_x}{k'_{-d}(k_x + k^0)} \; e^{-\Delta G_r^0/RT} \tag{43}$$

Now consider a series of closely related quenchers, so that $k'_d$, $k'_{-d}$, $k_x$ and $k^0$ can be considered constant. If the relation between $\Delta G_{12}^{**}$ and $\Delta G_r^0$ were that given by Marcus [Eq. (28)], a plot of log $k_q$ *vs.* $\Delta G_r^0$ should be parabolic as is that of Fig. 10. In particular, when $|-\Delta G_r^0|$ is larger than $2[\Delta G^{**}(^*A, A_1^+) + \Delta G^{**}(A_2, A_2^-)]$, *i.e.*, than $\lambda_{12}$, the quenching constant $k_q$ should decrease as the free energy change becomes more favorable (inverted region). Rehm and Weller[85, 86], however, in their electron-transfer quenching of aromatic hydrocarbon fluorescence did not find any evidence of a decrease of the rate constant up to $\Delta G_r^0 = -62$ kcal/mol, although the inverted region was expected at $-10$ kcal/mol. Thus, they assumed that $\Delta G_{12}^{**}$ is a monotonic function of $\Delta G_r^0$ [Eq. (44)]

$$\Delta G_{12}^{**} = \frac{\Delta G_r^0}{2} + \left[ \left(\frac{\Delta G_r^0}{2}\right)^2 + \left(\frac{\lambda_{12}}{4}\right)^2 \right]^{1/2} \tag{44}$$

Using this assumption, Eq. 42 can be further reduced to Eq. (45)

$$k_q = \frac{k'_d}{1 + \frac{k'_{-d}}{k^0}} \tag{45}$$

and a plot of log $k_q$ *vs.* $\Delta G_r^0$ looks like that of Fig. 15, where the value of the plateau for $\Delta G_r^0$ large and negative is given by Eq. (45) and the slope of the linear portion for $\Delta G_r^0$ large and positive is $-1/2 \cdot 3$ RT [Eq. (43)]. In the previous equations $\Delta G_r^0$ can be expressed in terms of redox potentials as in Eq. (46),

$$\Delta G_r^0 = E^0(A_1^+/^*A_1) - E^0(A_2/A_2^-) + w_p - w_r \tag{46}$$

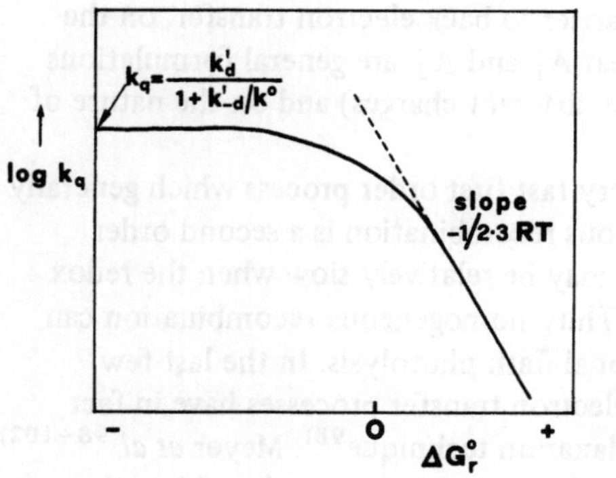


Fig. 15. Plot of the logarithm of the quenching constant *vs* the free energy change for the electron transfer step according to Eqs. (41), (43) and (45)

where the redox potential of the excited state can be approximately obtained from the redox potential of the ground state and the zero-zero spectroscopic energy of the excited state, [Eqs. (12) and (13)] as has been discussed in detail in Section 5.

The treatment given above accounts fairly well for the results obtained in the electron-transfer quenching of fluorescent aromatic molecules[85, 86], thionine and lumiflavin acid and basic triplet forms[93], duroquinone triplet state[94] and the lowest excited states of $Cr(bpy)_3^{3+}$ and $Ir(Me_2phen)_2Cl_2^+$ [54]. In all cases, no evidence of $k_q$ decrease was found in the inverted region. "Vestiges" (*i.e.*, a very small decrease in $k_q$) of the inverted region have however been found in the quenching of polypirydine ruthenium(II) complexes[87]. Further experimental and theoretical investigations are clearly needed in order to understand the details of the excited state electron-transfer reactions.

## 7. Electron Transfer Reactions Induced by Excited State Electron Transfer Quenching

Because of the difference in ground and excited state potentials, electron transfer quenching can lead to ground state redox products for which a thermal back electron transfer reaction is highly favored. This happens, for example, if the energy situation is that depicted in Fig. 14. As one can see from Fig. 13, the back electron transfer reaction can occur immediately in the successor complex obtained as the result of the electron transfer quenching. In this case, the back reaction is generally called cage recombination. Alternatively, if $k_s$ is competitive with $k_{-e}$ the successor complex can dissociate into the solvent separated oxidized and reduced species, whose back electron transfer reaction is called homogeneous recombination[d]. Of course, $A_1^+$ and/or $A_2^-$ can also disappear through other reaction paths (*e.g.*, reaction with the solvent) not shown in Fig. 13. The efficiency of separation between

[d] The spin problems related to geminate and homogeneous recombination of radical ion pairs in organic photochemistry have recently been discussed by Schulten *et al.*[95].

$A_1^+$ and $A_2^-$ will depend on the activation barrier to back electron transfer, on the charge type of the redox products (recall that $A_1^+$ and $A_2^-$ are general formulations for redox products which may actually have different charges) and on the nature of the medium[96, 97].

The cage recombination reaction is a very fast first order process which generally cannot be observed, whereas the homogeneous recombination is a second order process which, even if diffusion controlled, may be relatively slow when the redox products are formed in low concentration. Thus, homogeneous recombination can usually be observed using laser or conventional flash photolysis. In the last few years, the rate constants of many thermal electron transfer processes have in fact been obtained using flash photolysis as a relaxation technique[98]. Meyer *et al.*[98–102] have shown that besides measuring the back electron transfer reaction this technique can be used in a designed way for measuring rates of electron transfer processes involving species which do not participate in the photochemical electron transfer process. Consider, for example, the redox equilibrium

$$B + C \rightleftharpoons B^- + C^+ \tag{47}$$

and suppose it is strongly displaced to the left-hand side. If we take a molecule A whose excited state is able to reduce B and whose oxidized form is able to oxidize C, a flash photolysis experiment can yield the following reaction sequence:

$$A \xrightarrow[\text{flash}]{h\nu} {}^*A \tag{48}$$

$$^*A + B \longrightarrow A^+ + B^- \tag{49}$$

and

$$A^+ + C \longrightarrow A + C^+ \tag{50}$$

The result of this sequence is the formation of stoichiometric amounts of $B^-$ and $C^+$ as a consequence of A excitation. That is, A has been used as a "photocatalyst" for driving the reaction (47) in the non-spontaneous direction[99]:

$$B + C \underset{}{\overset{\text{light}}{\rightleftharpoons}} B^- + C^+ \tag{51}$$

The thermal back electron transfer reaction which restores the thermodynamic equilibrium can thus be studied. In principle, any redox reaction can be driven if the potentials for its component redox couples fall within the energy gap between the $A^+/{}^*A$ and $A^+/A$ couples. Analogous systems can be devised in which the oxidizing ability (instead of the reduxing ability) of the excited state is used[100].

As we have seen in Section 5, an excited state can be both a good reductant and a good oxidant. Thus, it can be unstable towards disproportionation:

$$2\ {}^*A \longrightarrow A^+ + A^- \tag{52}$$

The experimental conditions for reaction (52), however, are hard to obtain because it is difficult to achieve a sufficiently high concentration of excited states. Moreover, reaction (52) could be slow for kinetic reasons (Marcus "inverted" region, Section 6.1). It is interesting to note that reaction (52) can be driven by an appropriate modification of the above mentioned method[102]:

$$
\begin{aligned}
&2A \xrightarrow{2h\nu} 2\ {}^*A \\
&{}^*A + B \longrightarrow A^+ + B^- \\
&{}^*A + C \longrightarrow A^- + C^+ \\
\hline
&2A + B + C \xrightarrow{2h\nu} A^+ + A^- + B^- + C^+
\end{aligned}
\qquad (53)
$$

Finally, it should be noted that when the back electron transfer reaction between the primary products ($A_1^+$ and $A_2^-$ in Fig. 13) is sufficiently slow, the concentrations of these products attain appreciable steady-state values. In such a case, a photogalvanic effect can be observed in the system (Section 8).

## 8. Conversion of Light Energy into Chemical Energy

In the last few years there has been an increasing interest in solar energy utilization. Photochemical processes (and particularly photochemical electron transfer processes) are very suitable to convert light energy into more useful energy forms[103–106].

Direct conversion of solar energy into electrical energy can be obtained by means of photogalvanic cells, in which a photochemical reaction causes a change in the concentration of the electroactive species. The principles and applications of photogalvanic cells have been recently reviewed[107–110].

Any endoergonic photochemical reaction converts light energy into chemical energy. If the photoproducts are kinetically stable, they have to be considered "fuels" as they can be stored, transported and then converted to other chemical species with evolution of energy[111]. As we have seen in Section 5, electronic excitation increases the oxidation and reduction potentials of a molecule. Light absorption can thus drive a redox reaction in the non-spontaneous direction (Fig. 16).

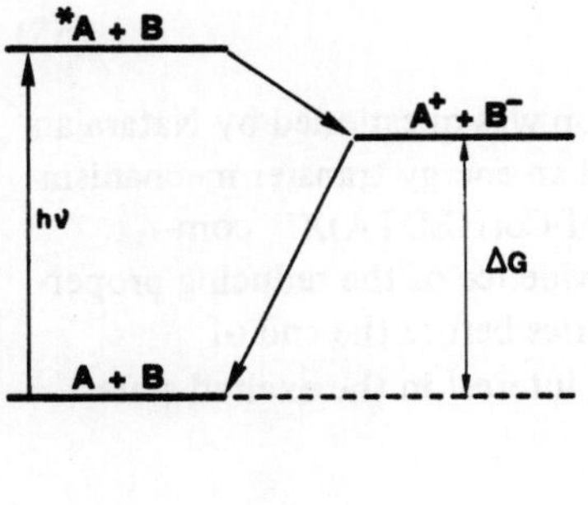


Fig. 16. Conversion of light energy into chemical energy by an electron-transfer reaction

Practically all the excited state energy can be converted into chemical energy in an outer sphere excited state reaction (Section 9.3). The successful development of redox systems capable of using light energy requires the prevention of back-reactions that return the primary photochemical products to the initial starting material form. Thus, in the scheme of Fig. 16 it is important that the thermal back electron transfer reaction be relatively slow so as to allow the build up (and eventually the storage) of the high energy products $A^+$ and $B^-$. These thermal electron transfer reactions can be treated in terms of the current electron transfer theories (Section 6.2) and in favorable cases their rates can be manipulated by systematic variations in A, B and the solvent. A more effective way of preventing the thermal back-reaction is by building into the system a degree of irreversibility, which is functionally equivalent to electrical rectification. This however introduces some energy loss. In this regard, reactions with solvent (especially, acid-base reactions) could be particularly useful. The most efficient means of preventing the back-reactions is the use of two or more phases to effect the separation of the primary electron transfer products. This could be done by appropriate interfacial or membrane systems, as happens in natural photosynthetic processes[112]. Progress in this direction, however, seems to be rather slow.

## 9. Complexes Containing 2,2′-Bipyridine, 1,10-Phenanthroline or Their Derivatives as Ligands

### 9.1. Introduction

In the last few years it has been found that the complexes containing aromatic molecules such as bipyridine, phenanthroline or their derivatives as ligands are particularly suitable for excited state electron transfer reactions. These complexes have octahedral structures and can be obtained with a variety of transition metal ions[113]. The best studied among these complexes is certainly $Ru(bpy)_3^{2+}$, but systematic investigations carried out in several laboratories are rapidly broadening our knowledge on this very interesting family of excited-state electron-transfer reagents.

Electron transfer quenching of $Ru(bpy)_3^{2+}$ was first proposed by Gafney and Adamson[114] in 1972. The quenchers used were pentaamine Co(III) complexes which are known to decompose upon one electron reduction:

$$^*Ru(bpy)_3^{2+} + Co(NH_3)_5Br^{2+} \longrightarrow Ru(bpy)_3^{3+} + Co(NH_3)_5Br^+ \qquad (54)$$

$$Co(NH_3)_5Br^+ \longrightarrow Co^{2+} + 5\,NH_3 + Br^- \qquad (55)$$

The interpretation given by Gafney and Adamson was questioned by Natarajan and Endicott[115, 116], who had previously suggested an energy transfer mechanism for the $Ru(bpy)_3^{2+}$ sensitized photodecomposition of $Co(HEDTA)X^-$ complexes[4, 117–119]. However, definite experimental evidence of the reducing properties of $^*Ru(bpy)_3^{2+}$ was obtained in several laboratories before the end of 1974[120–123]. Since then, there has been a growing interest in the excited-state redox reactions of $Ru(bpy)_3^{2+}$ and related systems, particularly because of their potential for the solar energy conversion problem.

Some important properties of the bpy and phen complexes are as follows:

1) they can be oxidized and reduced without disruption of the molecular structure[11, 124–127)];
2) their absorption spectra generally extend to the visible region with intense absorption bands[9, 10, 113)];
3) in their normal oxidation states they are generally stable in an aqueous solution at room temperature[113, 128)] and they do not undergo ligand photodissociation[30, 129) e)];
4) the lowest excited state is generally populated with unitary efficiency regardless of the excitation wavelength[13, 132–134)];
5) the Stokes shift between absorption to and emission from the lowest excited state is generally small[133, 135–138)];
6) the lowest excited state exhibits luminescence emission even in a fluid solution at room temperature[13, 139–141)];
7) the lifetime of the lowest excited state[13, 139–141)] is generally long enough to allow the participation of the excited state in bimolecular processes;
8) the relatively long lifetime of the lowest excited state allows the measurement of the excited state absorption spectrum[30, 134, 140)].

A selection of the available data concerning important properties of these complexes is given in Tables 1–4.

## 9.2. Spectroscopic Properties

The absorption and emission spectra of $Ru(bpy)_3^{2+}$ are shown in Fig. 17. The bands at 185 nm (not shown in the figure) and 285 nm have been assigned to LC $\pi \rightarrow \pi^*$

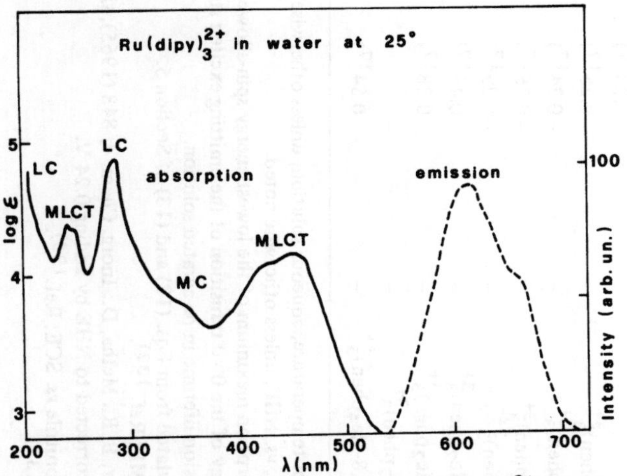


Fig. 17. Absorption and emission spectra of $Ru(bpy)_3^{2+}$ in aqueous solution at room temperature

e) However, ligand photosubstitution takes place in non-aqueous solvents[130, 131)].

Table 1. Spectroscopic and redox properties of polypyridine Cr(III) complexes[1]

| Complex, M | $E^{0\prime}(M^+/M)$[2] (V) | $E^{0\prime}(M/M^-)$[2] (V) | $\lambda_{max}^{abs}$[3] (nm) | $E_{0-0}(M-{}^*M)$[4] (eV) | $\tau$[5] (μs) | $^*E^{0\prime}(M^+/{}^*M)$[6] (V) | $^*E^{0\prime}({}^*M/M^-)$[6] (V) |
|---|---|---|---|---|---|---|---|
| $Cr(bpy)_3^{3+}$ | > 1.6[7] | −0.25[8] | 455[10] | 1.71[11] | 77[12] | > −0.1 | 1.46 |
| | | −0.26[9] | | 1.71[11] | 50[13] | | 1.45 |
| $Cr(4,4'\text{-}\phi_2bpy)_3^{3+}$ | – | – | 430[14] | 1.67[14] | 108[14] | – | – |
| $Cr(phen)_3^{3+}$ | – | −0.25[15] | 430[14] | 1.71[11] | 270[14] | – | 1.46 |
| | | −0.23[16] | | | | | |
| $Cr(5\text{-}Clphen)_3^{3+}$ | – | −0.25[17] | 464[14] | 1.70[14] | 68[14] | – | 1.45 |
| $Cr(5\text{-}Brphen)_3^{3+}$ | – | −0.24[17] | – | – | – | – | – |
| $Cr(5\text{-}Mephen)_3^{3+}$ | – | −0.36[17] | – | – | – | – | – |
| $Cr(5\text{-}\phi phen)_3^{3+}$ | – | −0.29[17] | – | – | – | – | – |
| $Cr(4,7\text{-}Me_2phen)_3^{3+}$ | – | −0.49[17] | 450[14] | 1.69[14] | 158[14] | – | 1.20 |
| $Cr(5,6\text{-}Me_2phen)_3^{3+}$ | – | −0.38[17] | – | – | – | – | – |
| $Cr(4,7\text{-}\phi_2phen)_3^{3+}$ | – | – | 490[14] | 1.67[14] | 240[14] | – | – |
| $Cr(3,4,7,8\text{-}Me_4phen)_3^{3+}$ | – | −0.54[17] | – | – | – | – | – |

1) Room temperature, aqueous solution, unless otherwise noted.
2) $E_{1/2}$, *vs.* NHE, unless otherwise noted.
3) Absorption maximum of the lowest energy spin-allowed transition.
4) Energy of the 0–0 transition of the emitting excited state.
5) Emission lifetime in de-aerated solution.
6) Calculated from Eqs. (12) and (13) of Section 5.
7) In DMF; Ref.[132].
8) Baker, B. R., Metha, D.: Inorg. Chem. *4*, 848 (1965); the value *vs.* SCE was corrected to NHE by adding 0.24 V.
9) Acetonitrile *vs.* SCE; Ref.[125].
10) Ref.[137].
11) Ref.[138].
12) Ref.[30].
13) Acetonitrile, air equilibrated; Ref.[54].
14) Ref.[141].
15) Soignet, D. M., Hargis, L. G.: Inorg. Chem. *14*, 941 (1975); the value *vs.* SCE was corrected to NHE by adding 0.24 V.
16) Hughes, M. C., Macero, D. J.: Inorg. Chem. *15*, 2040 (1976); acetonitrile *vs.* Ag/AgCl.
17) Hughes, M. C., Macero, D. J.: Inorg. Chem. *13*, 2739 (1974); in water-ethanol 50%.

Table 2. Spectroscopic and redox properties of polypyridine Ru(II) complexes[1]

| Complex, M[2] | $E^{0\prime}(M^+/M)$[3] (V) | $E^{0\prime}(M/M^-)$[3] (V) | $\lambda_{max}^{abs}$[4] (nm) | $E_{0-0}(M-{}^*M)$[5] (eV) | $\tau$[6] ($\mu s$) | $^*E^{0\prime}(M^+/{}^*M)$[7] (V) | $^*E^{0\prime}({}^*M/M^-)$[7] (V) |
|---|---|---|---|---|---|---|---|
| $Ru(bpy)_3^{2+}$ | 1.26[8] | −1.28[11,12] | 452[8] | 2.12[13] | 0.624[14] | −0.86 | 0.84 |
| | 1.29[9,10] | −1.35[10,12] | | 2.12[13] | 0.85[14,15] | −0.83 | 0.77 |
| $Ru(4,4'\text{-}Me_2bpy)_3^{2+}$ | 1.10[8] | −1.37[16] | 460[8] | 2.04[17] | 0.33[8] | −0.94 | 0.67 |
| $Ru(4,4'\text{-}\phi_2bpy)_3^{2+}$ | 1.17[8] | – | 474[8] | 2.02[13] | 0.67[8] | −0.85 | – |
| $Ru(4,4'\text{-}R1_2bpy)_3^{2+}$ | 1.59[9,18] | −0.90[9,18] | 466[15,18] | 2.23[15,18] | 2.39[15,18] | −0.64 | 1.33 |
| $Ru(4,4'\text{-}R2_2bpy)_3^{2+}$ | 1.53[9,18] | −0.91[9,18] | 467[15,18] | 2.21[15,18] | 2.21[15,18] | −0.68 | 1.30 |
| $Ru(4,4'\text{-}R3_2bpy)_3^{2+}$ | 1.55[9,18] | −0.95[9,18] | 467[15,18] | 2.21[15,18] | 1.93[15,18] | −0.66 | 1.26 |
| $Ru(4,4'\text{-}R4_2bpy)_3^{2+}$ | – | – | 468[15,18] | 2.23[15,18] | 2.19[15,18] | – | – |
| $Ru(4,4'\text{-}R5_2bpy)_3^{2+}$ | – | – | 467[15,18] | 2.22[15,18] | 2.21[15,18] | – | – |
| $Ru(bpy)_2(4,4'\text{-}R6_2bpy)^{2+}$ | 1.34[9,18] | −0.99[9,18] | 468[18,19] | 2.12[18,19] | 2.00[18,19] | −0.78 | 1.13 |
| $Ru(bpy)(4,4'\text{-}R6_2bpy)_2^{2+}$ | – | – | 468[18,19] | 2.17[18,19] | 2.10[18,19] | – | – |
| $Ru(4,4'\text{-}R6_2bpy)_3^{2+}$ | – | – | 468[18,19] | 2.20[18,19] | 2.14[18,19] | – | – |
| $Ru(bpy)_2(CN)_2$ | 1.09[20] | −1.30[20] | – | 2.15[21] | 0.22[21] | −1.30 | 0.61 |
| $Ru(phen)_3^{2+}$ | 1.26[8] | −1.36[11,12] | 447[8] | 2.18[13] | 0.81[14] | −0.92 | 0.82 |
| | 1.29[9,10] | −1.43[9,12] | | 2.18[13] | 0.50[14,15] | −0.89 | 0.75 |
| $Ru(5\text{-}Mephen)_3^{2+}$ | 1.23[8] | (−1.31)[16] | 450[8] | 2.15[22] | 1.33[8] | −0.92 | (0.84) |
| $Ru(5,6\text{-}Me_2phen)_3^{2+}$ | 1.20[8] | (−1.34)[16] | 453[8] | 2.14[22] | 1.81[8] | −0.94 | (0.80) |
| $Ru(4,7\text{-}Me_2phen)_3^{2+}$ | 1.09[8] | −1.47[16] | 445[8] | 2.12[22] | 1.74[8] | −1.03 | 0.65 |
| $Ru(5\text{-}Clphen)_3^{2+}$ | 1.36[8] | −1.15[16] | 447[8] | 2.15[22] | 0.94[8] | −0.79 | 1.00 |
| $Ru(5\text{-}Brphen)_3^{2+}$ | 1.37[8] | – | 448[8] | 2.15[22] | 1.04[8] | −0.78 | – |
| $Ru(5\text{-}NO_2phen)_3^{2+}$ | 1.46[8] | – | 449[8] | 2.15[22] | ⩽0.005[8] | −0.69 | – |
| $Ru(5\text{-}\phi phen)_3^{2+}$ | 1.26[8] | – | 448[8] | 2.15[22] | 1.29[8] | −0.89 | – |
| $Ru(4,7\text{-}\phi_2phen)_3^{2+}$ | 1.20[8] | – | 460[8] | 2.08[13] | 4.68[8] | −0.88 | – |

Table 2. (continued)

| Complex, M[2] | $E^{0'}(M^+/M)$[3] (V) | $E^{0'}(M/M^-)$[3] (V) | $\lambda_{max}^{abs}$[4] (nm) | $E_{0-0}(M-{}^*M)$[5] (eV) | $\tau$[6] ($\mu s$) | $^*E^{0'}(M^+/{}^*M)$[7] (V) | $^*E^{0'}({}^*M/M^-)$[7] (V) |
|---|---|---|---|---|---|---|---|
| $Ru(3,5,6,8\text{-}Me_4phen)_3^{2+}$ | 1.09[8] | – | 440[8] | 2.15[22] | 2.22[8] | −1.06 | – |
| $Ru(3,4,7,8\text{-}Me_4phen)_3^{2+}$ | 1.02[8] | – | 438[8] | 2.15[22] | 1.39[8] | −1.13 | – |
| $Ru(phen)_2(CN)_2$ | 1.12[23] | −1.79[23] | – | 2.19[21] | 1.09[21] | −1.07 | 0.40 |
| $Ru(terpy)_2^{2+}$ | 1.25[9, 24] | −1.40[9, 24] | 473[8] | – | ⩽0.005[8] | – | – |
| $Ru(TPTZ)_2^{2+}$ | 1.49[9, 24] | −0.81[9, 24] | 501[8] | – | ⩽0.005[8] | – | – |
| $Ru(terpy)(bpy)(NH_3)^{2+}$ | – | – | 462[14, 15] | – | 0.43[14] | – | – |

[1]) Room temperature, aqueous solution, unless otherwise noted.
[2]) See list of symbols.
[3]) $E_{1/2}$ or $E_p$−0.03, *vs.* NHE, unless otherwise noted.
[4]) Absorption maximum of the lowest energy spin-allowed transition.
[5]) Energy of the 0–0 transition of the emitting excited state.
[6]) Emission lifetime in de-aerated solution.
[7]) Calculated from Eqs. (12) and (13) of Section 5.
[8]) Ref.[156].
[9]) Acetonitrile *vs.* SCE.
[10]) Ref.[53].
[11]) The value in $CH_3CN$ *vs.* SCE was corrected to $H_2O$ *vs.* NHE by adding 0.07 V.
[12]) Ref.[163].
[13]) Ref.[136].
[14]) Ref.[173].
[15]) Acetonitrile.
[16]) Ref.[140].
[17]) Calculated from the reduction potentials of ground and excited states reported in Ref.[156].
[18]) Ref.[185].
[19]) Isobutyronitrile.
[20]) DMF *vs.* NHE, Ref.[132].
[21]) DMF, Ref.[162].
[22]) Values calculated from Ref.[156] adding 0.1 eV to the corrected emission maximum. This difference is approximately that experimentally found for $Ru(bpy)_3^{2+}$ and $Ru(phen)_3^{2+}$.
[23]) In DMF, from Ref.[126]; the value *vs.* SCE was corrected to NHE by adding 0.24 V.
[24]) Ref.[11].

Table 3. Spectroscopic and redox properties of polypyridine Os(II) complexes[1]

| Complex, M | $E^{0'}(M^+/M)$[2] (V) | $E^{0'}(M/M^-)$[2] (V) | $\lambda_{max}^{abs}$[3] (nm) | $E_{0-0}(M-{}^*M)$[4] (eV) | $\tau$[5] (ns) | $^*E^{0'}(M^+/{}^*M)$[6] (V) | $^*E^{0'}({}^*M/M^-)$[6] (V) |
|---|---|---|---|---|---|---|---|
| $Os(bpy)_3^{2+}$ | 0.83[7] | −1.18[7] | 640[8] | 1.85[9] | 19[10] | −1.02 | 0.67 |
| | | | | | 49.0[9] | | |
| $Os(4,4'\text{-}Me_2bpy)_3^{2+}$ | (0.72)[10] | −1.31[10] | – | 1.78[10] | ~ 9[10] | (−1.06)[10] | 0.50[10] |
| $Os(phen)_3^{2+}$ | 0.82[10] | (−1.21)[10] | 650[11] | 1.78[10, 12] | 84[7] | −0.96 | (0.57) |
| | | | | | 183[9] | | |
| $Os(5\text{-}Clphen)_3^{2+}$ | 0.93[10] | −1.09[10] | – | 1.78[10] | 78[10] | −0.85[10] | 0.69[10] |
| $Os(5\text{-}Mephen)_3^{2+}$ | 0.79[10] | (−1.24)[10] | – | 1.78[10] | 69[10] | −0.99[10] | (0.54)[10] |
| $Os(5,6\text{-}Me_2phen)_3^{2+}$ | 0.75[10] | (−1.28)[10] | – | 1.78[10] | 63[10] | −1.03[10] | (0.50)[10] |
| $Os(phen)_2[4,7\text{-}(SO_3\phi)_2phen]$ | – | – | – | 14.8[9] | 93[9] | – | – |
| $Os(phen)_2(4,7\text{-}\phi_2phen)^{2+}$ | – | – | – | 15.0[9] | 212[9] | – | – |

1) Room temperature, aqueous solution, unless otherwise noted.
2) $E_{1/2}$ *vs.* NHE, unless otherwise noted.
3) Absorption maximum of the lowest energy spin-allowed transition.
4) Energy of the 0–0 transition of the emitting excited state.
5) Emission lifetime in de-aerated solution.
6) Calculated from Eqs. (12) and (13) of Section 5.
7) Ref.[160].
8) Bryant, G. M., Fergusson, J. E., Powell, H. K. J.: Austr. J. Chem. *24*, 257 (1971); in MeOH.
9) Ref.[168]; in MeOH.
10) Ref.[140].
11) A. Juris, unpublished results.
12) A value of 1.90 eV is reported in Ref.[168].

Table 4. Spectroscopic and redox properties of polypyridine Ir(III) complexes[1])

| Complex, M | $E^{0\prime}(M^+/M)$[2]) (V) | $E^{0\prime}(M/M^-)$[2]) (V) | $\lambda_{max}^{abs}$[3]) (nm) | $E_{0-0}(M-{}^*M)$[4]) (eV) | $\tau$[5]) (μs) | $^*E^{0\prime}(M^+/{}^*M)$[6]) (V) | $^*E^{0\prime}({}^*M/M^-)$[6]) (V) |
|---|---|---|---|---|---|---|---|
| $Ir(bpy)_3^{3+}$ | 2.17[7]) | −0.76[7]) | ~440[8]) | 2.81[9]) | 2.4[9]) | −0.64 | 2.05 |
| $Ir(bpy)_2Cl_2^+$ | 2.11[7]) | −1.03[7]) | 448[10]) | 2.62[11]) | – | −0.51 | 1.59 |
| $Ir(phen)_3^{3+}$ | – | – | – | 2.79[9]) | 2.9[9]) | – | – |
| $Ir(phen)_2Cl_2^+$ | – | −1.12[12]) | 390[13]) | 2.67[14]) | 0.33[13, 15]) | – | 1.55[16]) |
| $Ir(5,6\text{-}Me_2phen)_2Cl_2^+$ | – | −1.16[12]) | 370[13]) | 2.54[14]) | 0.92[13,15])<br>0.49[17]) | – | 1.38[16]) |

[1]) Room temperature, aqueous solution, unless otherwise noted.
[2]) $E_{1/2}$ *vs.* NHE, unless otherwise noted.
[3]) Absorption maximum of the lowest energy spin-allowed transition.
[4]) Energy of the 0–0 transition of the emitting excited state.
[5]) Emission lifetime in de-aerated solution.
[6]) Calculated from Eqs. (12) and (13) of Section 5.
[7]) Ref. [127]); the value in $CH_3CN$ *vs.* SCE was corrected to $H_2O$ *vs.* NHE by adding 0.07 V.
[8]) Flynn, C. M. Jr., Demas, N. J.: J. Am. Chem. Soc. *96*, 1959 (1974).
[9]) Ref. [168]), in MeOH.
[10]) R. Ballardini, unpublished results.
[11]) Ref.[135]); in 4:1 EtOH-MeOH.
[12]) Ref.[126]); in DMF *vs.* SCE.
[13]) Ref.[46]).
[14]) Ref.[135]); in 1:1 water-glycole.
[15]) Acetonitrile.
[16]) In DMF *vs.* SCE.
[17]) Ref.[54]); acetonitrile, air equilibrated.

transitions by comparison with the spectrum of protonated bipyridine[142]. The two remaining intense bands at 240 and 450 nm have been assigned to MLCT d → π* transitions, each one showing two vibronic components resolved at room temperature[142]. The shoulders at 322 and 344 nm could probably be MC transitions. The assignments of the long-wavelength absorption tail above 500 nm has been the object of a long dispute. Its interpretation as a "spin-forbidden" MLCT transition was finally found to be consistent with all the experimental data[142, 143]. Theoretically, the lowest empty MO of this complex is expected[11] to be a π* LC orbital instead of a σ* MC orbital as in Fig. 1. The orbital nature of the emitting state is now definitely established as MLCT[142, 143] but the spin label, which is usually taken as a triplet, cannot really be defined because of the dominant role played by spin-orbit coupling in the MLCT excited states[17]. Other investigations on the photophysics of $Ru(bpy)_3^{2+}$ have been reported[129, 144–151].

A change in the nature of the central metal may cause a change in the orbital nature of the excited state, which is MC for $Cr(bpy)_3^{3+}$ and LC for $Ir(bpy)_3^{3+}$. MLCT excited states (such as those of $Ru(bpy)_3^{2+}$ and $Os(bpy)_3^{2+}$) and LC excited states [*e.g.*, $Ir(bpy)_3^{3+}$ and $Ir(5,6\text{-}Me_2phen_2)_2Cl_2^+$] involve promotion of an electron to highly delocalized π* ligand orbitals and thus they are not expected to be strongly distorted with respect to the ground state geometry (Section 5). The relatively small values of the Stokes shifts for the $Ru(bpy)_3^{2+}$ [133, 136] and $Ir(5,6\text{-}Me_2phen_2)_2Cl_2^+$ [135] are in agreement with such an expectation. MC excited states involving promotion to $\sigma_M^*$ orbitals are expected to be strongly distorted. The lowest MC excited state of Cr(III) complexes, as mentioned in Section 5, is obtained from intraconfiguration transitions within the partially filled $\pi_M(t_{2g})$ orbitals[2, 3] and thus it is not expected to be distorted. This is experimentally proved for $Cr(bpy)_3^{3+}$ and $Cr(phen)_3^{3+}$, the Stokes shift of which is zero[137, 138].

As we can see from Tables 1–4, a change in the nature of the central metal causes profound changes in all the excited-state properties. Substitution of phenanthroline for bipyridine does not affect the orbital nature of the lowest excited state and only causes an increase in the excited state lifetime. However, Crosby[12] has shown that the orbital nature of the lowest excited state can also be changed by substitution of bpy or phen with other ligands. For example, as we have seen in Section 2, the lowest excited state is LC for $Ir(phen)_3^{3+}$ [13] MLCT for $Ir(phen)_2Cl_2^+$ [12] and MC for $Ir(phen)Cl_4^-$ [12]. Ligand substituents do not change the nature of the lowest excited state except when the energy separation between different states is very small[135]. However, they can cause noticeable changes in the excited state lifetime and excited state redox potentials (Tables 1–4). Of course, the possibility of changing the excited state properties by appropriate choice of metal, ligands and ligand substituents is most useful for systematic studies as well as for possible applications.

The absorption spectra of $^*Ru(bpy)_3^{2+}$ [134, 140], $^*Os(bpy)_3^{2+}$ [140] and $^*Cr(bpy)_3^{3+}$ [30] have been determined by the flash-photolysis technique. The first two complexes, whose lowest excited states have MLCT orbital nature, exhibit similar absorption spectra with intense bands at 360 and 430 nm for $^*Ru(bpy)_3^{2+}$ and at 360 and 460 nm for $^*Os(bpy)_3^{2+}$. The similarity with the spectrum of the $bpy^-$ radical anion[152] probably cannot be taken as evidence against the delocalization of the excited electron over the three bpy ligands, which has been proposed by Crosby[62, 153] on the basis of the spectroscopic behavior of bpy-phen mixed complexes. For $^*Cr(bpy)_3^{3+}$ the absorption spectrum shows several relatively low intensity bands which have been tentatively assigned to MC transitions[30].

## 9.3. Redox Properties of the Excited States

As we have seen in Section 5, for excited states which are not very distorted with respect to the ground state the reduction potentials of the $M^+/^*M$ and $^*M/M^-$ couples are approximately given by the reduction potentials of the corresponding couples involving the ground state molecule minus and, respectively, plus the one-electron potential corresponding to the excited state spectroscopic energy. For $Ru(bpy)_3^{2+}$ (Fig. 18), as the reduction potential of the ground state is –1.28 V

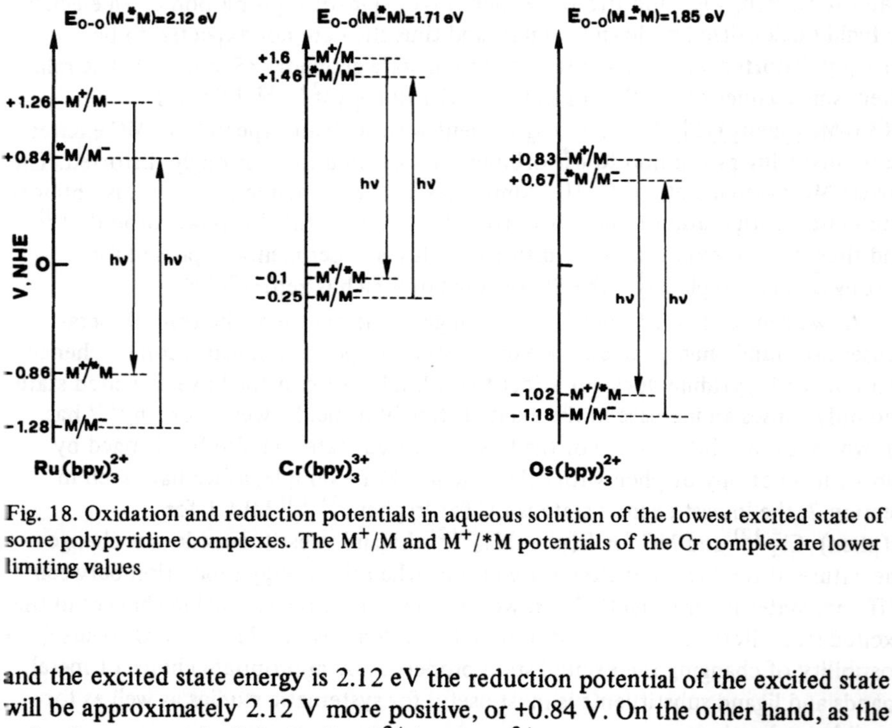


Fig. 18. Oxidation and reduction potentials in aqueous solution of the lowest excited state of some polypyridine complexes. The $M^+/M$ and $M^+/^*M$ potentials of the Cr complex are lower limiting values

and the excited state energy is 2.12 eV the reduction potential of the excited state will be approximately 2.12 V more positive, or +0.84 V. On the other hand, as the reduction potential of the $Ru(bpy)_3^{3+}/Ru(bpy)_3^{2+}$ couple is +1.26 V, the reduction potential of the $Ru(bpy)_3^{3+}/^*Ru(bpy)_3^{2+}$ couple will be 2.12 V more negative, or –0.86 V. Thus the excited state of this complex is a moderately strong oxidant and at the same time a strong reductant. If we take another complex, we will have different excited state redox properties because of different ground state redox properties and different excited state energies. For example, $^*Cr(bpy)_3^{3+}$ (Fig. 18) is a very strong oxidant but a poor reductant.

The redox potential of an excited state can also be evaluated from kinetic measurements. As we have seen in Section 6.2 the quenching constant for the electron transfer quenching of an excited state by a homogeneous family of quenchers having variable oxidation or reduction potentials is a function of the free energy change [Eqs. (41) and (44)]. Assuming reasonable values for the rate constants that appear in Eq. (41), by using the best fitting procedure[54, 121] it is possible to find the value of the quencher potential that corresponds to $\Delta G_r^0 = 0$; so that the excited state potential can be straightforwardly calculated. In this way, the following values have been obtained (acetonitrile solutions, *vs.* SCE): $*Cr(bpy)_3^{3+}/Cr(bpy)_3^{2+}$, $E_{1/2} \simeq +1.3$ V[54]; $*Ru(bpy)_3^{2+}/Ru(bpy)_3^{+}$, $E_{1/2} \simeq +0.7$ V[54]; $Ru(bpy)_3^{3+}/*Ru(bpy)_3^{2+}$, $E_{1/2} = -0.83$ V[121]; $*Ir(5,6\text{-}Me_2phen)_2Cl_2^{+}/Ir(5,6\text{-}Me_2phen)_2Cl_2$, $E_{1/2} \simeq +1.25$ V[54]. Considering the large experimental errors, these values are in reasonable agreement with those (Tables 1, 2, 4) calculated following the procedure discussed in Section 5.

### 9.4. Quenching Processes

In bimolecular processes, the lowest excited state of these complexes can act as an energy donor [Eq. (56)], electron acceptor [Eq. (57)] and electron donor [Eq. (58)].

$$*M + Q \xrightarrow{\text{en. tr.}} M + *Q \qquad \text{energy transfer} \tag{56}$$

$$*M + Q \xrightarrow{\text{red.}} M^- + Q^+ \qquad \text{reductive quenching} \tag{57}$$

$$*M + Q \xrightarrow{\text{ox}} M^+ + Q^- \qquad \text{oxidative quenching} \tag{58}$$

From the thermodynamic point of view, the ability of an excited state to be involved in energy transfer processes is related to the excited state energy; and the ability to be involved in electron transfer processes is related to the excited state reduction and oxidation potentials. Among the three complexes of Fig. 18 the excited Os complex is the best electron donor; the excited Cr complex is the best electron acceptor and the excited Ru complex is the best energy donor. It is interesting to note that the excited Ru complex is at the same time, a good energy donor, a good electron acceptor and a good electron donor. The quenching process(es) taking place will depend on the specific properties of the excited state and the quencher. For example, in the

$$*Ru(bpy)_3^{2+} + Tl^{3+} \xrightarrow{\text{en.tr.}}\!\!\!\!\!/\;\; Ru(bpy)_3^{2+} + *Tl^{3+} \qquad (-2.12\ \text{eV};\ > +4.5\ \text{eV}) \tag{59}$$

$$*Ru(bpy)_3^{2+} + Tl^{3+} \xrightarrow{\text{red}}\!\!\!\!\!/\;\; Ru(bpy)_3^{+} + Tl^{4+} \qquad (-0.84\ \text{eV};\ > +2\ \text{eV}) \tag{60}$$

$$\overset{-0.86\ \text{eV}}{} \quad *Ru(bpy)_3^{2+} + Tl^{3+} \xrightarrow{\text{ox}} Ru(bpy)_3^{3+} + Tl^{2+} \quad \underset{-0.35\ \text{eV}}{} \tag{61}$$

quenching of $*Ru(bpy)_3^{2+}$ by $Tl^{3+}$ [123], energy transfer [Eq. (59)] and reductive electron transfer [Eq. (60)] are thermodynamically not allowed, whereas oxidative electron transfer [Eq. (61)] is thermodynamically very favorable. Thus, there is little doubt that the quenching takes place by Eq. (61). In other cases however, more than one quenching mechanism may be thermodynamically allowed. With $Fe^{3+}$ as a quencher[121, 123, 154–156] both energy transfer [Eq. (62)] and oxidative electron transfer [Eq. (64)] are thermodynamically allowed. In continuous irradiation experiments[123] the quenching takes place without

$$\overset{-2.12\ \text{eV}}{} \quad *Ru(bpy)_3^{2+} + Fe^{3+} \xrightarrow{\text{en.tr.}} Ru(bpy)_3^{2+} + *Fe^{3+} \quad \underset{+1.6\ \text{eV}}{} \tag{62}$$

$$\overset{-0.84\ \text{eV}}{} \quad *Ru(bpy)_3^{2+} + Fe^{3+} \xrightarrow{\text{red}}\!\!\!\!\!/\ \ Ru(bpy)_3^{+} + Fe^{4+} \quad \underset{>+1\ \text{eV}}{} \tag{63}$$

$$\overset{-0.86\ \text{eV}}{} \quad *Ru(bpy)_3^{2+} + Fe^{3+} \xrightarrow{\text{ox}} Ru(bpy)_3^{3+} + Fe^{2+} \quad \underset{-0.73\ \text{eV}}{} \tag{64}$$

any permanent chemical change. On the other hand, $*Fe^{3+}$ is known to be a non-emitting excited state. Thus, from continuous irradiation experiments one cannot say whether the quenching occurs *via* reaction (62) or (64). Flash photolysis experiments[121] and steady state measurements[155] have shown that $Ru(bpy)_3^{3+}$ and $Fe^{2+}$ are transiently produced, at first sight suggesting that the quenching occurs *via* reaction (64). A closer examination of this system, however, shows that the formation of $Ru(bpy)_3^{3+}$ and $Fe^{2+}$ in the quenching process does not necessarily discriminate between energy transfer and oxidative electron transfer quenching mechanisms[4, 157]. This is evident from an energy diagram like that of Fig. 19, where we can see that energy transfer (step 3) would lead to the lowest excited state of $Fe^{3+}$ which, lying at 12.6 kK (1.6 eV) above the ground state[15], should have a reduction potential of about 2.3 V. This means that step 3 could be followed by step 4, which leads to the same transient products as does the primary electron transfer quenching (step 2). Only a very detailed analysis of the various rate constants has led to the conclusion that electron transfer (step 2) is probably the major quenching mechanism in this system[156]. Another system in which discrimination between thermodynamically allowed energy and electron transfer has been successful is $*Ru(bpy)_3^{2+} + Cr(bpy)_3^{3+}$ [158], where electron transfer quenching is again

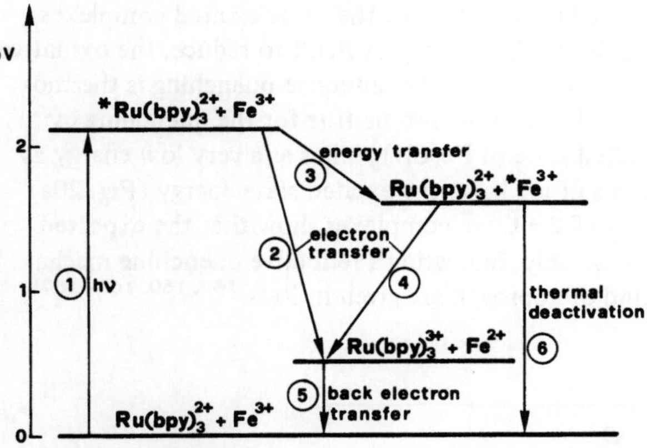


Fig. 19. Energy diagram showing the possible deactivation paths for the *Ru(bpy)$_3^{2+}$ + Fe$^{3+}$ system

predominant. Of course, just as the formation of electron transfer products is not *per se* a proof of an electron transfer quenching mechanism, so the formation of an excited quencher is not *per se* a proof of an energy transfer quenching because the primary electron transfer quenching products $M^-$ and $Q^+$ (or $M^+$ and $Q^-$) could undergo subsequent reaction to yield M and *Q.

Useful information on the nature of the quenching mechanism can be obtained from the comparison of the quenching constants of related systems. As we have seen in Section 6.2, it is expected that for homogeneous families of excited states and/or quenchers when an electron transfer mechanism is operative the quenching constant increases as the process becomes thermodynamically more favorable, until a plateau region is reached which corresponds to the diffusion controlled value[f]. This is also the case for quenching *via* energy transfer[20, 159]. Thus, plots of log $k_q$ *vs* the free energy change for the various quenching processes can often elucidate the nature of the quenching mechanism.

Consider, for example, the quenching of *Cr(bpy)$_3^{3+}$, *Ru(bpy)$_3^{2+}$ and *Os(bpy)$_3^{2+}$ by Fe(CN)$_6^{4-}$ [160]. In principle, the quenching may occur, as we have seen before, by energy transfer or by reductive or oxidative electron transfer:

$$*M + Fe(CN)_6^{4-} \xrightarrow{\text{en.tr.}} M + *Fe(CN)_6^{4-} \quad ? \tag{65}$$

$$*M + Fe(CN)_6^{4-} \xrightarrow{\text{red}} M^- + Fe(CN)_6^{3-} \quad +0.44\ \text{eV} \tag{66}$$

$$*M + Fe(CN)_6^{4-} \xrightarrow{\text{ox}} M^+ + Fe(CN)_6^{5-} \quad > +1.7\ \text{eV} \tag{67}$$

f) Whether or not $k_q$ decreases when the reaction becomes highly exoergonic (Marcus inverted region) is still the object of discussion (Section 9.6).

The excited state energy and the redox potentials of the three excited complexes are given in Tables 1, 2 and 4. As $Fe(CN)_6^{4-}$ is very difficult to reduce, the oxidative quenching [Eq. (67)] can be readily discarded. The reductive quenching is thermodynamically allowed in all cases and this could also be true for the quenching by energy transfer if the lowest excited state of $Fe(CN)_6^{4-}$ lies at a very low energy as some people believe[161]. The plots of log $k_q$ *vs* the excited state energy (Fig. 20a) and the reduction potential (20b) of the three complexes show that the expected correlation holds in the second case only, indicating a reductive quenching mechanism. Other examples of this kind of approach are given in Refs.[140, 160, 162–167].

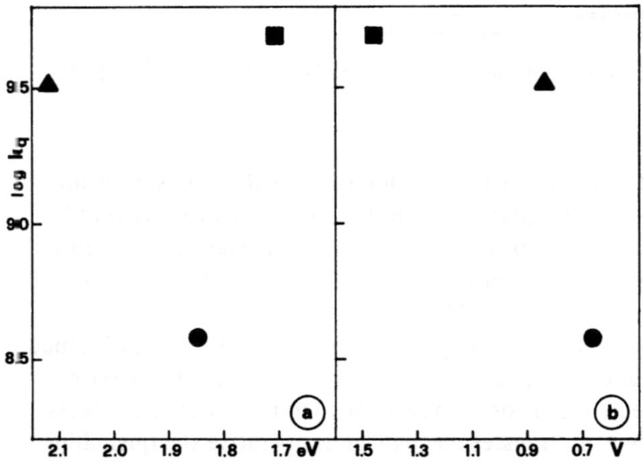


Fig. 20. Plots of log $k_q$ *vs* the excited state energy (a) and the reduction potential (b) of *M in the quenching of *M by $Fe(CN)_6^{4-}$. $*Cr(bpy)_3^{3+}$, ■, $*Ru(bpy)_3^{2+}$, ▲, $*Os(bpy)_3^{2+}$, ● [160]

Energy *vs* electron transfer quenching of $*Ru(bpy)_3^{2+}$ and $*Ru(phen)_3^{2+}$ by a number of inorganic species is discussed in Ref. 161. Oxygen quenching of excited $Ru(bpy)_3^{2+}$ and other polypyridine complexes yields singlet oxygen[13, 168], but the reaction mechanism presents some complicated aspects in acid medium[169] and an electron transfer quenching mechanism (producing $O_2^-$), followed by a back electron transfer reaction leading to $^1O_2$ is also plausible[13, 154, 156].

Usually the existence, and thus the quenching, of an excited state is proved by means of luminescence measurements. However, relatively long lived excited states can also exist in the absence of a detectable emission and it may be possible to observe their electron transfer reactions by observing products in flash experiments[7, 98]. It may also be possible to estimate excited-state lifetimes by observing product yields as a function of quencher concentration. This extends considerably the number of systems that can be investigated. It is also interesting to note that an electrochemical method has been devised for measuring the lifetime of non-emitting excited states[170].

Other interesting studies on the quenching of polypyridine complexes are reported in Refs.[171–180].

## 9.5. Conversion of Light Energy into Chemical Energy

As we have seen in Section 8, excited-state electron-transfer reactions can be used for the conversion of light energy into chemical energy. Most of the polypyridine complexes show intense absorption bands in the visible region (see, for examples, Fig. 17) and thus they are particularly suitable for solar energy conversion.

As we have seen in Section 9.3, practically all the excited energy of these complexes can be used as free energy in electron transfer processes. The amount of excited state energy actually converted will depend on the redox potential of the reaction partner. For example, reaction (64) converts 0.53 eV, *i.e.*, 25% of the excited state energy, whereas reaction (68)[158] converts 1.51 eV, (71%). This last reaction is a part of a very

$$\overbrace{{}^*Ru(bpy)_3^{2+} + Cr(bpy)_3^{3+} \xrightarrow{ox}}^{-0.86\ eV} \underbrace{Ru(bpy)_3^{3+} + Cr(bpy)_3^{2+}}_{+0.25\ eV} \tag{68}$$

interesting photochemical system which is obtained when an aqueous solution containing both $Ru(bpy)_3^{2+}$ and $Cr(bpy)_3^{3+}$ is irradiated (Fig. 21)[158, 181]. In the dark, the equilibrium is strongly displaced towards the left-hand side because the ground state $Ru(bpy)_3^{2+}$ is a poor reductant and the ground state $Cr(bpy)_3^{3+}$ is a poor oxidant. Depending on the excitation wavelength, the light can be absorbed by the Ru or Cr complex. In the first case one obtains the excited Ru complex,

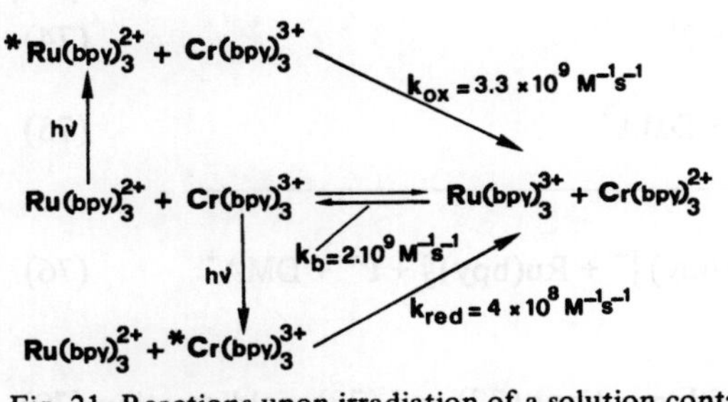


Fig. 21. Reactions upon irradiation of a solution containing $Ru(bpy)_3^{2+}$ and $Cr(bpy)_3^{3+}$ [158, 181]

which is a good reductant, and thus reacts with $Cr(bpy)_3^{3+}$ to give $Ru(bpy)_3^{3+}$ and $Cr(bpy)_3^{2+}$. On the other hand, light absorption by $Cr(bpy)_3^{3+}$ leads to $^*Cr(bpy)_3^{3+}$, which is an excellent oxidant, and thus reacts with $Ru(bpy)_3^{2+}$ to give again $Ru(bpy)_3^{3+}$ and $Cr(bpy)_3^{2+}$. Thus, the same products are obtained in this system, irrespective of the absorbing species. The compounds so obtained are a strong oxidant and a strong reductant, *i.e.*, high energy species whose reaction liberates 1.51 eV. Unfortunately, as it will be discussed later, this converted energy is only transiently stored because these species rapidly undergo a back electron-transfer reaction which regenerates the starting materials.

It is also possible to design systems in which these complexes are used as photocatalysts to drive a reaction in the non-spontaneous direction. An example is shown by Eqs. (69)–(72), where $P^{2+}$ is the *paraquat* dication and the solvent is acetonitrile[98–100]

$$Ru(bpy)_3^{2+} \xrightarrow{h\nu} {}^*Ru(bpy)_3^{2+} \tag{69}$$

$$^*Ru(bpy)_3^{2+} + P^{2+} \longrightarrow Ru(bpy)_3^{3+} + P^+ \tag{70}$$

$$Ru(bpy)_3^{3+} + N\phi_3 \longrightarrow Ru(bpy)_3^{2+} + N\phi_3^+ \tag{71}$$

---

$$N\phi_3 + P^{2+} \xrightarrow{h\nu} N\phi_3^+ + P^+ \tag{72}$$

$$\Delta G = +1.42 \text{ eV}$$

Using high concentrations of $N\phi_3$, reaction (71) prevails over the back electron transfer reaction between $Ru(bpy)_3^{3+}$ and $P^+$ and the effect of the experiment is to use light absorbed by $Ru(bpy)_3^{2+}$ for driving the reaction between $N\phi_3$ and $P^{2+}$ (reaction 72) in the non-spontaneous direction. As $^*Ru(bpy)_3^{2+}$ is both a good oxidant and a good reductant (Fig. 18), it is also possible to design experiments in which both processes take place simultaneously in the same solution. This can be obtained, for example, by using appropriate concentrations of paraquat ($P^{2+}$) and N,N-dimethylaniline (DMA) as quenchers[7, 98, 102]:

$$2\ Ru(bpy)_3^{2+} \xrightarrow{+2h\nu} 2\ {}^*Ru(bpy)_3^{2+} \tag{73}$$

$$^*Ru(bpy)_3^{2+} + P^{2+} \longrightarrow Ru(bpy)_3^{3+} + P^+ \tag{74}$$

$$^*Ru(bpy)_3^{2+} + DMA \longrightarrow Ru(bpy)_3^{+} + DMA^+ \tag{75}$$

---

$$2\ Ru(bpy)_3^{2+} + P^{2+} + DMA \xrightarrow{2h\nu} Ru(bpy)_3^{3+} + Ru(bpy)_3^{+} + P^+ + DMA^+ \tag{76}$$

$$\Delta G = +3.84 \text{ eV}$$

Reactions (73)–(75) taken together with reactions (77) and (78) yield reaction (79), where light is again catalytically converted into chemical energy by $Ru(bpy)_3^{2+}$. On the other hand, reactions (73)–(75) plus reaction (80) represent the catalyzed, light induced disproportionation of $Ru(bpy)_3^{2+}$ into $Ru(bpy)_3^{3+}$ and $Ru(bpy)_3^{+}$ [reaction (81)][g].

$$Ru(bpy)_3^{3+} + DMA \longrightarrow Ru(bpy)_3^{2+} + DMA^+ \tag{77}$$

$$Ru(bpy)_3^{+} + P^{2+} \longrightarrow Ru(bpy)_3^{2+} + P^+ \tag{78}$$

---

g) The recently reported "triplet-triplet" annhilation of $^*Ru(bpy)_3^{2+}$ in a micelle solution[182] is probably a direct, light induced disproportionation of this complex [Eq. (52)].

$$2\,DMA + 2\,P^{2+} \xrightarrow[\Delta G = +2.40\ eV]{+2h\nu} 2DMA^{+} + 2P^{+} \qquad (79)$$

$$P^{+} + DMA^{+} \longrightarrow P^{2+} + DMA \qquad (80)$$

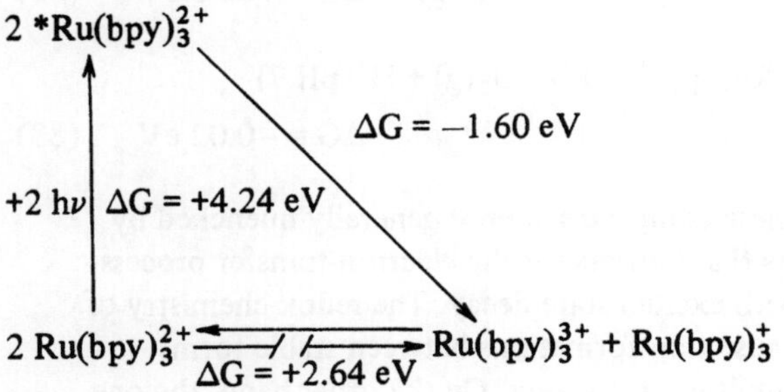


(81)

While the *conversion* of light energy into chemical energy is relatively easy, the storage of the converted energy is generally prevented by rapid return of the products to the unexcited starting materials in a succeeding electron transfer event. As a matter of fact, in all systems examined above the high energy products [*e.g.*, $Ru(bpy)_3^{3+}$ and $Fe^{2+}$ in reaction (64), $Ru(bpy)_3^{3+}$ and $Cr(bpy)_3^{2+}$ in reaction (68), $N\phi_3^{+}$ and $P^{+}$ in reaction (72), etc.] are only obtained as transients and their formation can only be observed by flash techniques (incidentally, it should be noted that the reduced forms of the complexes can usually be observed more easily than the oxidized forms because of their high intensity absorption bands in the visible)[100, 158, 163, 175, 183]. In order to avoid immediate energy dissipation in the back electron-transfer step, one should slow down the rate of this step and/or scavenge one of the high energy partners formed in the electron transfer quenching process to give products which will have a lower energy content but high kinetic stability. Whitten and co-workers[184, 185] have recently shown that for "hydrophobic" derivatives of $Ru(bpy)_3^{2+}$ (see Table 2) the shielding of the reactive core by the hydrophobic sheath may impart a selectivity to electron transfer reactions. Thus, whereas the quenching of $^*Ru(bpy)_3^{2+}$ by DMA or $N\phi_3$ is followed by fast back electron-transfer reactions so that the high energy $Ru(bpy)_3^{+}$ species can only be obtained transiently, in the quenching of the hydrophobic derivatives of $Ru(bpy)_3^{2+}$ the corresponding Ru(I) species is obtained as a permanent product in a dry acetonitrile solution. Evidently, with the hydrophobic complexes the energy wasting back-reaction has been retarded sufficiently to allow other reactions of the amine radical cation[184, 185].

Another approach to the utilization of the free energy change caused by light absorption in redox systems like those described above is that based on the photogalvanic effect[108, 109]. A rather large photogalvanic potential (~ 0.18 V) has been observed by Lin and Sutin[154] for the $Ru(bpy)_3^{2+}$ $-Fe^{3+}$ system which may offer some advantages (*e.g.*, a better reversibility) with respect to the well known iron-thionine system[2, 109, 186–188].

The best way to utilize solar energy is, of course, the photocatalized splitting of water[103–106]. A consideration of reduction potentials indicates that excited states of complexes such as $Ru(bpy)_3^{2+}$ should be able to reduce water as hydrogen ion to hydrogen gas over a wide pH range, and at the same time to oxidize water at pH 7:

$$^*Ru(bpy)_3^{2+} + H^+(pH\ 7) \longrightarrow Ru(bpy)_3^{3+} + 1/2 \cdot H_2(g) \quad \Delta G = -0.45\ eV \qquad (82)$$

$$^*Ru(bpy)_3^{2+} + 1/2 \cdot H_2O \longrightarrow Ru(bpy)_3^{+} + 1/4 \cdot O_2(g) + H^+(pH\ 7) \quad \Delta G = -0.02\ eV \qquad (83)$$

The fact that excited states of these complexes are not generally quenched by water or acid in solution suggests that barriers for the electron-transfer process render it too slow to compete with excited state decay. The redox chemistry of water is in fact complicated because transformations between stable forms ($H_2O \rightarrow H_2 + 1/2 \cdot O_2$) involve multielectron steps. On the other hand, the one-electron reduction of $H^+$ to H and the one-electron oxidation of $H_2O$ to $H^+$ + OH require much more energy than that available to the excited complex[103]. In order to circumvent these difficulties, one can devise cycles in which the excited state undergoes a simple electron transfer reaction to yield a reduced or oxidized quencher which is able to react with water. For example, one could take into con-

$$Ru(bpy)_3^{2+} \xrightarrow{h\nu} {}^*Ru(bpy)_3^{2+} \qquad (84)$$

$$^*Ru(bpy)_3^{2+} + Eu^{3+} \longrightarrow Ru(bpy)_3^{3+} + Eu^{2+} \qquad (85)$$

$$Eu^{2+} + H_2O \longrightarrow Eu^{3+} + 1/2 \cdot H_2 + OH^- \qquad (86)$$

$$Ru(bpy)_3^{3+} + OH^- \longrightarrow Ru(bpy)_3^{2+} + 1/4 \cdot O_2 + 1/2 \cdot H_2O \qquad (87)$$

---

$$1/2 \cdot H_2O \xrightarrow{h\nu} 1/2 \cdot H_2 + 1/4 \cdot O_2 \qquad (88)$$

sideration the cycle constituted by reactions (84)–(87). Unfortunately, although each reaction is known to occur separately[111, 154, 156, 189], the system does not work because the back electron transfer reaction between $Ru(bpy)_3^{3+}$ and $Eu^{2+}$ prevails over reactions (86) and (87). Other cycles have been discussed by Meyer[7, 98], but no practical progress has so far been made.

Whitten *et al.*[190, 191] reported the splitting of water into hydrogen and oxygen upon irradiation of monolayer assemblies of surfactant derivatives of $Ru(bpy)_3^{2+}$. Subsequent investigations, however, have shown that more highly purified samples of these complexes are inactive toward catalytic light-induced cleavage of water and that these systems are strongly affected by factors which often lead to irreproducible results[192–194].

The use of an electrochemical cell to effect the photodecomposition of water using visible light has been proposed by Creutz and Sutin[189]. This cell consists of an *n*-type semiconductor connected to a platinum electrode in a pH 9 buffer containing $Ru(bpy)_3^{2+}$. The excited Ru complex injects an electron in the conduction band of the semiconductor (see also Ref.[195]) thus forming $Ru(bpy)_3^{3+}$, which is then reduced by $OH^-$ to the starting complex (and $O_2$). The electrons released by the $*Ru(bpy)_3^{2+}$ flow from the semiconductor electrode through the external circuit to the platinum electrode, where they effect the reduction of water (or hydrogen ion) to $H_2$. A quantitative extension[196] of that study has shown that the possibility of such systems is limited by stringent but probably not prohibitive conditions.

Other approaches to the use of these complexes in practical systems for the utilization of solar energy include (i) the creation of appropriate chemically linked systems (for example, $Ru(bpy)_2(py)^{2+}$ units bound to a poly-vinyl-pyridine resin) with the aim of obtaining "cooperative" effects in the electron transfer reactions[197] and (ii) the stabilization of an excited state or a primary redox product for a time long enough to allow the absorption of a second photon.

## 9.6. Kinetic Aspects

As we have seen in Section 6.1, the bimolecular quenching constant of an excited-state electron-transfer reaction [Eq. (41)] depends on the free activation energy and the free energy change of the electron transfer step. According to Marcus[71, 73, 78], the relationship between these two quantities is given by Eq. (28) [or Eq. (25)], whereas Weller[85, 86] has suggested the relationship given by Eq. (44). For the quenching of an excited state by a homogeneous series of quenchers having variable redox potential, the reorganizational parameter $\lambda_{12}$ is constant and thus the free activation energy [Eqs. (28) or (44)] is only a function of the free energy change, which in its turn depends on the redox potentials of the excited state and the quenchers. According to Eq. (28), the plot of log $k_q$ *vs.* the free energy change [Eq. (41)] should be parabolic (see, for example, Fig. 10), whereas according to Eq. (44) $k_q$ should increase as the reaction becomes more and more thermodynamically favored until it reaches a limiting value (Fig. 15) which is maintained even for very large and negative free energy changes. The two plots almost coincide for positive or slightly negative free energy changes[86], but they are very different for large and negative free energy changes (Marcus inverted region).

Ballardini *et al.*[54] have recently carried out a systematic investigation on the quenching of $*Cr(bpy)_3^{3+}$, $*Ru(bpy)_3^{2+}$ and $*Ir(5,6\text{-}Me_2phen)_2Cl_2^+$ by a series of methoxybenzenes, aromatic amines and aliphatic amines in an acetonitrile solution. Flash photolysis experiments showed that the quenching process takes place with formation of the reduced complex and the one-electron oxidation product of the quencher, indicating an electron-transfer quenching mechanism. Using the excited state reduction potentials given in Tables 1, 2 and 4, which have been obtained from Eq. (13), the plots of log $k_q$ *vs.* the free energy change of the electron transfer reaction (Fig. 22) show no evidence of the Marcus inverted region, even for the Cr and Ir complexes whose reactions with the most reducing quenchers are highly exoergonic. In such cases, however, an excited state of the reduced complex or

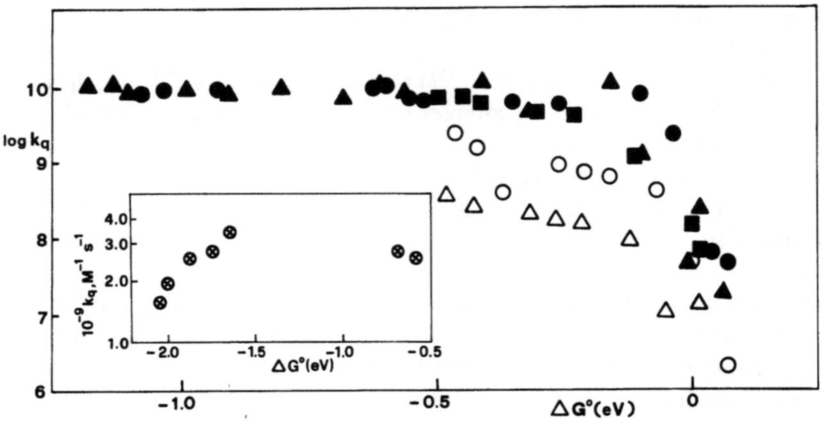


Fig. 22. Plot of log $k_q$ *vs.* $\Delta G^0$ for the electron transfer quenching of $^*Cr(bpy)_3^{3+}$ (▲, △), $^*Ru(bpy)_3^{2+}$ (■) and $^*Ir(5,6\text{-}Me_2phen)_2Cl_2^+$ (●, ○) by aromatic quenchers (full points), and aliphatic amines (empty points)[54]. The insert reports the data for the quenching of $^*Ru(bpy)_3^{2+}$ and $^*Ru(4,4'\text{-}Me_2bpy)_3^{2+}$ by $Cr(bpy)_3^{3+}$, $Ru(bpy)_3^{3+}$, $Os(bpy)_3^{3+}$ and $Ru(4,4'\text{-}Me_2bpy)_3^{3+}$[87]. Note the expanded ordinate scale in the insert

oxidized quencher could have been obtained (Fig. 11), which would imply a high rate constant even according to the Marcus theory. On the other hand, theoretical treatments[76, 88, 90–92] have recently been proposed which predict a much smaller (if any) decrease of $k_q$ in the inverted region.

From Fig. 22 it is also apparent that aliphatic amines behave differently from aromatic amines or methoxybenzenes. This is not surprising for three reasons[54]: (i) the oxidation of aliphatic amines is an irreversible process and the $E_{1/2}$ values used to obtain $\Delta G^0$ may not represent thermodynamic quantities; (ii) the reorganization energy is expected to be larger and thus the reaction slower for aliphatic than for aromatic amines because the electron to be transferred is much more localized, which implies a much more extended geometrical change upon oxidation; (iii) owing to the greater localization, steric factors can slow down the reaction rate of aliphatic amines. Slower electron transfer quenching by aliphatic amines compared to aromatic amines has also been observed for triplet methylene blue[198, 199].

In another systematic investigation, Creutz and Sutin[87] studied the quenching of $^*Ru(bpy)_3^{2+}$ and $^*Ru(4,4'\text{-}Me_2bpy)_3^{2+}$ by $Cr(bpy)_3^{3+}$, $Os(bpy)_3^{3+}$, $Ru(bpy)_3^{3+}$ and $Ru(4,4'\text{-}Me_2bpy)_3^{3+}$ and found a small decrease of $k_q$ with increasing $|-\Delta G^0|$ (Fig. 22, insert) in qualitative agreement with the Marcus theory. The values of $k_q$ obtained, however, are many orders of magnitude above those predicted by the Marcus theory, as expected on the basis of the recent theoretical treatments mentioned above.

The experiments of Ballardini *et al.*[54] have also shown that the reorganizational parameter for the quenching of $^*Cr(bpy)_3^{3+}$, $^*Ru(bpy)_3^{2+}$ and $^*Ir(5,6\text{-}Me_2phen)_2Cl_2^+$ by aromatic quenchers is small, suggesting that the self-exchange electron-transfer reactions of these excited states are as fast as the ground state self-exchange reactions[87, 101, 164]. This is in agreement with the small Stokes shift values[133, 135–138)], which indicate that the excited states are quite similar in size, shape and solvation to the ground states.

It is interesting to compare the ground and excited-state electron-exchange reactions in terms of orbital occupancy. For the Ru complex[140)], whose excited state is MLCT, the exchange between $*Ru(bpy)_3^{2+}$ and $Ru(bpy)_3^{3+}$ [Eq. (91)]

$$\underset{(\pi_M)^6(\pi_L^*)^0}{Ru(bpy)_3^{2+}} + \underset{(\pi_M)^5(\pi_L^*)^0}{Ru(bpy)_3^{3+}} \underset{}{\overset{k_{II,\,III}}{\rightleftharpoons}} \underset{(\pi_M)^5(\pi_L^*)^0}{Ru(bpy)_3^{3+}} + \underset{(\pi_M)^6(\pi_L^*)^0}{Ru(bpy)_3^{2+}} \tag{89}$$

$$\underset{(\pi_M)^6(\pi_L^*)^0}{Ru(bpy)_3^{2+}} + \underset{(\pi_M)^6(\pi_L^*)^1}{Ru(bpy)_3^{+}} \overset{k_{II,\,I}}{\rightleftharpoons} \underset{(\pi_M)^6(\pi_L^*)^1}{Ru(bpy)_3^{+}} + \underset{(\pi_M)^6(\pi_L^*)^0}{Ru(bpy)_3^{2+}} \tag{90}$$

$$\underset{(\pi_M)^5(\pi_L^*)^1}{*Ru(bpy)_3^{2+}} + \underset{(\pi_M)^5(\pi_L^*)^0}{Ru(bpy)_3^{3+}} \overset{*k_{II,III}}{\rightleftharpoons} \underset{(\pi_M)^5(\pi_L^*)^0}{Ru(bpy)_3^{3+}} + \underset{(\pi_M)^5(\pi_L^*)^1}{*Ru(bpy)_3^{2+}} \tag{91}$$

$$\underset{(\pi_M)^5(\pi_L^*)^1}{*Ru(bpy)_3^{2+}} + \underset{(\pi_M)^6(\pi_L^*)^1}{Ru(bpy)_3^{+}} \overset{*k_{II,\,I}}{\rightleftharpoons} \underset{(\pi_M)^6(\pi_L^*)^1}{Ru(bpy)_3^{+}} + \underset{(\pi_M)^5(\pi_L^*)^1}{*Ru(bpy)_3^{2+}} \tag{92}$$

involves the transfer of a $\pi_L^*$ electron, as does the exchange between $Ru(bpy)_3^{2+}$ and $Ru(bpy)_3^{+}$ [Eq. (90)]. On the other hand, the exchange between $*Ru(bpy)_3^{2+}$ and $Ru(bpy)_3^{+}$ [Eq. (92)] involves the transfer of a $\pi_M$ electron, as does the exchange between $Ru(bpy)_3^{2+}$ and $Ru(bpy)_3^{3+}$ [Eq. (89)]. In both cases no drastic change is expected in the metal-ligand bonds or in the ligands because the $\pi_M$ orbitals are almost non-bonding and the $\pi_L^*$ orbitals are delocalized. An activation barrier, however, arises from outer-sphere reorganization, owing to the change in the ionic charge. Because of the large size of the reacting ions, this barrier is small and in media of high ionic strength, where the electrostatic repulsion term, $w_r$, is negligible, the reactions have rates near the diffusion-controlled limit[101)]. For $Cr(bpy)_3^{3+}$, whose excited state is intraconfigurational MC (Section 2), both the ground and the excited state reductive self-exchange processes [Eqs. (93) and (94)] involve the transfer of a non-bonding $\pi_M$

$$\underset{(\pi_M)^3(\pi_L^*)^0}{Cr(bpy)_3^{3+}} + \underset{(\pi_M)^4(\pi_L^*)^0}{Cr(bpy)_3^{2+}} \overset{k_{III,\,II}}{\rightleftharpoons} \underset{(\pi_M)^4(\pi_L^*)^0}{Cr(bpy)_3^{2+}} + \underset{(\pi_M)^3(\pi_L^*)^0}{Cr(bpy)_3^{3+}} \tag{93}$$

$$\underset{(\pi_M)^3(\pi_L^*)^0}{*Cr(bpy)_3^{3+}} + \underset{(\pi_M)^4(\pi_L^*)^0}{Cr(bpy)_3^{2+}} \overset{*k_{III,\,II}}{\rightleftharpoons} \underset{(\pi_M)^4(\pi_L^*)^0}{Cr(bpy)_3^{2+}} + \underset{(\pi_M)^3(\pi_L^*)^0}{*Cr(bpy)_3^{3+}} \tag{94}$$

electron and are thus expected to be fast. The inferences drawn in Ref.[200)] concerning the rates of self-exchange of excited $Ru(bpy)_3^{2+}$ and $Cr(bpy)_3^{3+}$ are theoretically wrong because they are based on wrong Stokes shift values. On the other hand, application of the cross Marcus relation [Eq. (33)] can yield wrong values if the quantities used are affected by large experimental errors or if factors not included in the equation (*e.g.*, orbital and spin factors) play an important role. For example, the oxidation of ground state $Ru(bpy)_3^{2+}$ [Eq. (89)] involves $\pi_M$ orbitals, whereas the corresponding excited state reaction [Eq. (91)] involves $\pi_L^*$ orbitals. This difference may be more or less important depending on the nature of the reaction partner (see also Ref.[163]).

Of course, when the excited state is distorted with respect to the ground state, the excited-state self-exchange reaction can be faster or slower than the ground state self-exchange depending on the type of distortion (Section 6.2).

## 10. Uranyl Ion

### 10.1. Excited State Properties

The uranyl ion luminesces in fluid solutions at room temperature[2, 201, 202], thus providing a tool for the study of bimolecular excited state processes. In several cases, however, this study is complicated by the fact that the uranyl ion forms complexes with a variety of chemical species, so that it is often difficult to distinguish between intramolecular photochemical processes involving uranyl ion-ligand complexes and intermolecular photochemical processes involving reaction between an electronically excited $UO_2^{2+}$ species and the substrate[2, 201, 202].

Assignment of the electronic excited states involved in the absorption spectrum of $UO_2^{2+}$ is still controversial[2, 201]. The lowest energy absorption region (20.5–30.0 kK) has been alternatively attributed to triplet[203–206] or singlet[207, 208] excited states, although probably a spin label is not justified because of extensive spin orbit coupling[209, 210]. From the orbital point of view, the lowest energy bands are due to electron transfer from the two highest filled molecular orbitals (mainly consisting of oxygen 2p orbitals) to the empty uranium 5f shell[210, 211]. The lowest excited state of $UO_2^{2+}$ lies at 20.5 kK (2.54 eV)[202] and exhibits a characteristic luminescence, which in water decays with first order kinetics ($\tau = 6.6\ \mu s$)[202, 212]. The lifetime is affected by counter ions[202, 213], $UO_2^{2+}$ concentration[212, 214] and temperature[212, 214]. Flash-photolysis experiments have shown that the lowest excited state possesses absorption in the visible and near IR regions[212, 215–218], which may be used to monitor excited state bimolecular processes.

The standard reduction potential of the $UO_2^{2+}/UO_2^+$ couple is +0.06 V[219]; the oxidation potential of $UO_2^{2+}$ is unknown but is expected to be extremely high. Thus, the lowest excited state of $UO_2^{2+}$, $*UO_2^{2+}$, is expected to be a very good oxidant [Eq. (95)] and a very poor reductant (see Section 5).

$$E^0(*UO_2^{2+}/UO_2^+) \simeq E^0(UO_2^{2+}/UO_2^+) + E_{0-0} = 0.06 + 2.54 = 2.6\ V \qquad (95)$$

Photooxidation by $UO_2^{2+}$ has been demonstrated for a variety of organic and inorganic species. We will only review those systems in which an intermolecular mechanism is thought to be operative. Photooxidation by $UO_2^{2+}$ *via* an intermolecular mechanism can be simply schematized as in Eq. (96):

$$*UO_2^{2+} + Q \longrightarrow UO_2^{+} + Q^{+} \tag{96}$$

If Q bears a hydrogen atom, then it may lose it in the form of a proton in a step concerted with (96) to give a neutral free radical (Eq. 97):

$$*UO_2^{2+} + QH \longrightarrow UO_2^{+} + Q + H^{+} \tag{97}$$

$UO_2^{+}$ is usually considered to disproportionate[220–223], although it may be reoxidized to $UO_2^{2+}$ by a free radical or by molecular oxygen, in both of which cases no reduction of $UO_2^{2+}$ is apparent.

The formation of a $*(U_2O_4H^{4+})$ exciplex has recently been suggested to occur in $10^{-2}$M $HNO_3$ solution where dual luminescence was observed[224].

The photoacceleration of the U(VI)–U(IV) electron exchange involves the lowest excited state of $UO_2^{2+}$ and proceeds through the formation of an U(V) intermediate[223]. From this study, the quantum yield of formation of the lowest excited state of $UO_2^{2+}$ upon irradiation in the UV absorption band (300–330 nm) appeared to be ~ 0.5.

## 10.2. Quenching by Organic Molecules

The quenching of $*UO_2^{2+}$ emission by a number of alcohols has been found to be correlated with the photooxidation quantum yields[215, 225–227]. Large isotopic effects were found for the quenching by deuterated methanol, isopropanol and cyclohexanol, indicating that photooxidation occurs by a process of abstraction of an H atom bonded to carbon from a $-CH_2OH$ group[215, 218, 228]; however, H atom abstraction from the OH group has also been proposed for $CH_3OH$[229]. ESR studies at low temperature have confirmed that H atom abstraction from a carbon atom adjacent to an activating group such as –OH, –CO–, –CN, etc., is the most common mechanism of $*UO_2^{2+}$ oxidation of organic molecules, although C–C cleavage also (or only) occurs in several cases[230]. Photooxidation of lactic acid takes place by both intramolecular and intermolecular mechanisms with formation of acetaldehyde and pyruvic acid, respectively[231]. $*UO_2^{2+}$ photooxidation of acetaldehyde[232] and benzaldehyde[233] yields biacetyl and benzil. Theoretical analysis[234] of hydrogen abstraction in the $*UO_2^{2+}$ quenching by aliphatic compounds suggests that this reaction possesses considerable charge-transfer character. Hydrogen abstraction can also be the main route for deactivation of $*UO_2^{2+}$ in water at room temperature[212, 234]. The quenching of $*UO_2^{2+}$ by aromatic hydrocarbons[227, 235] and unsaturated compounds[235] takes place without the formation of permanent products. As the logarithm of the quenching constant correlates with the ionization potential of the hydrocarbons[227], the formation of a charge-transfer (exciplex) intermediate was suggested[227, 235]. Formation of exciplexes has also been assumed in the quenching of $*UO_2^{2+}$ by aromatic amines, phenols and other molecules[216, 224].

With some of these quenchers, formation of $UO_2^+$ and/or of the radical cation of the quencher was also observed in flash experiments[216, 217, 220]. However, it is not clear[214] how much of this proceeds from an intermolecular route, since $UO_2^{2+}$ is known to complex, for example, with aromatic amines.

### 10.3. Quenching by Metal Cations

Quenching of $^*UO_2^{2+}$ by metal cations has been studied by several authors[214, 236, 237]. Some of the results obtained have been collected in Table 5. It is evident that the absolute magnitudes of the quenching constants and also the order of the efficiencies of the various quenchers given by the two authors differ greatly. Since the different lifetimes of $^*UO_2^{2+}$ under the experimental conditions used can only introduce a proportionality constant between the two columns of $k_{SV}$ in Table 5, other factors

Table 5. Quenching of $^*UO_2^{2+}$ by metal ions

| Quencher | Stern-Volmer constant, $k_{SV}$, $M^{-1}$ Ref.[214] [1] | Ref.[237] [2] | $^*Q$ [3], kK | $E^0(Q^+/Q)$ [4], V |
|---|---|---|---|---|
| $Ag^+$ | 4350 | $5.2 \times 10^5$ | 44.7 | 1.99 |
| $Ba^{2+}$ | $<5 \times 10^{-3}$ | – | ≫ 20.5 | – |
| $Ce^{3+}$ | 0.38 | – | 28.6 [5] | 1.46 |
| $Co^{2+}$ | 12.1 | $3.7 \times 10^3$ | 8.1 | 1.84 |
| $Cu^{2+}$ | 3.36 | $1.1 \times 10^4$ | 9.4 | – |
| $Fe^{2+}$ | 833 | $2.6 \times 10^5$ | 8.3 | 0.43 |
| $Fe^{3+}$ | ≪ 67 | $3.2 \times 10^2$ | 12.6 | – |
| $Mn^{2+}$ | 4.25 | $5.6 \times 10^3$ | 18.9 | 1.51 |
| $Ni^{2+}$ | 6.9 | $1.5 \times 10^2$ | 8.5 | – |
| $Pb^{2+}$ | 40 | $1.8 \times 10^5$ | ≫ 20.5 | – |

[1]) $HNO_3$, pH 2–2.5, $[UO_2^{2+}] = 0.02$, room temperature.
[2]) $H_3PO_4$ 0.6 M, $[UO_2^{2+}] = 10^{-3}$, 28°.
[3]) Lowest excited state of the quencher, from Ref.[15] unless otherwise noted.
[4]) *vs.* NHE; Ref.[219].
[5]) Matthews, R. W., Sworski, T. J.: J. Phys. Chem. *79*, 681 (1975).

must be responsible for the observed differences. Presumably, $^*UO_2^{2+}$ and/or Q were present as different species under the different experimental conditions used. For these systems, static quenching *via* formation of a non-emissive complex in the ground state is unlikely. On the other hand, quenching by external heavy-atom effect or spin-catalyzed deactivation is also unlikely, in view of the fact that $^*UO_2^{2+}$ already contains a heavy atom and has a poorly defined spin number. Energy and electron transfer thus remain the most plausible quenching mechanisms[h]. Comparison of the

[h]) Quenching *via* exciplex formation has also been discussed very recently[238–241].

excited state energy (20.5 kK) and the excited state reduction potential (+2.6 V) of $UO_2^{2+}$ with the energy of the lowest excited state and the oxidation potential of the quencher (Table 5) shows that for several quenchers only electron transfer is energetically allowed. In flash photolysis experiments, a transient absorption assigned to $Mn^{3+}$ was observed for the $*UO_2^{2+}-Mn^{2+}$ system[214)]. With few exceptions, the quenching constants correlate with the ionization potentials[214, 237)] or with the oxidation potentials (Table 5), indicating that electron transfer quenching may dominate even when energy transfer is energetically allowed. In this regard, it is a particularly significant the fact that $Fe^{2+}$ is a much better quencher than $Co^{2+}$. It is also apparent that the ionic charge plays, as expected, an important role.

## 10.4. Quenching by Halide Ions

The quenching of uranyl luminescence by halide ions in solutions has been investigated by many workers[2, 201, 202)]. Quantitative results reported in recent papers have been collected in Table 6. The quenching constant decreases in the order

Table 6. Quenching of $*UO_2^{2+}$ by halide and halide-like ions[1)]

| Quencher | Quenching Constant, $k_q$, $M^{-1}s^{-1}$<br>1 M $HClO_4$ | <br>1 M $H_3PO_4$ | $E^0(X/X^-)$, V[2)] |
|---|---|---|---|
| $F^-$ | – | $(6 \times 10^4)$ [3)] | 3.6[4)] |
| $CN^-$ | – | $(3.1 \times 10^4)$ [3)] | 2.8 |
| $Cl^-$ | $1.7 \times 10^9$ | $8.0 \times 10^7$ | 2.6 |
| $Br^-$ | $4.8 \times 10^9$ | $8.5 \times 10^8$ | 2.07 |
| $SCN^-$ | – | $(1.8 \times 10^9)$ [3)] | – |
| $I^-$ | $6.9 \times 10^9$ | $2.1 \times 10^9$ | 1.42 |

1) 25°, $[UO_2^{2+}] = 0.01$; from Ref.[213)], unless otherwise noted.
2) *vs.* NHE, from Ref.[242)].
3) Evaluated from Ref. [237)] using $\tau = 150\ \mu s$[213)].
4) From Ref.[243)].

$I^- > Br^- > Cl^- \gg F^-$, which is the reverse of the order of the relative stabilities of the complexes of these anions with $UO_2^{2+}$ [237)]. This shows that static quenching is unlikely. More recently, it has been found that the quenching constants obtained from luminescence intensity measurements are the same as those obtained from luminescence lifetime measurements, which demonstrates a dynamic quenching mechanism[213)]. An almost linear correlation was found for log $k_q$ *vs.* $E^0(1/2X_2/X^-)$, suggesting an electron transfer mechanism[237)]. However, it should be noted that, in order to obtain a meaningful correlation between log $k_q$ and the free energy change on the electron transfer step the $E^0(X/X^-)$ values should be used. Comparison between the $*UO_2^{2+}/UO_2^+$ reduction potential (+2.6 V) and the $X/X^-$ reduction potentials (Table 6) shows that electron transfer quenching is thermodynamically allowed for

$I^-$, $Br^-$ and $Cl^-$, whereas it is forbidden for $F^-$ [213]. This prediction is in agreement with the values of the quenching constants reported in Table 6. Analysis of the quenching constants according to the Marcus equation (neglecting the work terms) indicates that the reorganizational parameter of $*UO_2^{2+}$ is higher in $H_3PO_4$ than in $HClO_4$ and that the reorganizational parameter of the halide ions increases from $Cl^-$ to $I^-$ [213]. This last finding was attributed to steric reasons[213]. A flash photolysis investigation has shown that $X_2^-$ radical ions (X = I, NCS, Br) are transiently formed upon irradiation of solutions containing $UO_2^{2+}$ and $X^-$ [244]. Similarly, a broad absorption assigned to $CO_3^-$ was observed for solutions containing $UO_2^{2+}$ and $NaHCO_3$, but whether the reaction occurs *via* intermolecular mechanism was unclear[244].

## 11. Formation and Reaction of Solvated Electrons

### 11.1. Introduction

Photoelectron production [Eq. (98)] is an important photochemical process for several transition metal complexes[2, 245]:

$$*A \longrightarrow A^+ + e_s^- \tag{98}$$

In principle, this is the simplest electron transfer reaction of an excited state. In practice, complications arise because of the very peculiar nature of the solvated electron[246, 247]. The way in which the electron separates from its parent molecule and the fate of the separated electron have been the object of much discussion.

Photoelectron ejection is for some molecules connected with the reaction of a solvated electron with other species, as it is known that in several cases this reaction leads to the formation of electronically excited states [Eq. (99)]. Whether reaction (99) or (100) prevails, when reaction (100) is very exoergonic, is of course important for a discussion of Marcus or anti-Marcus behavior (Section 6).

$$B + e_s^- \longrightarrow *B^- \tag{99}$$

$$B + e_s^- \longrightarrow B^- \tag{100}$$

### 11.2. Photoelectron Production

Although the formation of hydrated electrons upon UV irradiation of $Fe(CN)_6^{4-}$ was first reported in 1963[248], in 1970 there was still a substantial disagreement on the quantum yield of this reaction as well as on its relation with the spectroscopic excited states[2]. It is now established that the quantum yield of hydrated electron formation is 0.67 ± 0.03 for 254 nm excitation[249, 250], and that it decreases at longer wavelengths ($\Phi(e^-_{aq}) = 0.10$ at 313 and $< 0.02$ at 365 nm)[250] and increases at shorter wavelengths to reach a limiting value of ~ 0.9 below 228 nm[250]. In the case of simple inorganic anions, it is known that the formation of hydrated electrons from the excited state is connected with the charge-transfer-to-solvent (CTTS) nature of the transition[251]. In the case of $Fe(CN)_6^{4-}$, the UV absorption spectrum has some CTTS character at 270 nm but it is dominated by a MLCT band at 218 nm and MC bands at 270 and 322 nm[250]. According to Shirom and Stein[250, 252, 253], the hydrated electron formation occurs through the CTTS excited state which is reached either by direct excitation or by internal conversion from the intramolecular excited states below 313 nm. The nearly linear dependence of $\Phi(e^-_{aq})$ on the photon energy and the fact that $\Phi(e^-_{aq})$ at 228 nm is higher than $\Phi(e^-_{aq})$ at 254 nm (although the CTTS state is reached directly at 254 nm and *via* internal conversion at 228 nm) were taken as indications that the hydrated electrons are formed with higher yields from higher excited CTTS levels[250, 253]. Experiments in 10 N NaOH glasses have later suggested that the ejection of electrons may occur from different electronic states depending on the excitation wavelength and that the electrons ejected from a CTTS state have higher reactivity with a scavenger than electrons ejected from intramolecular states[254]. The quantum yields of solvated electrons from $Fe(CN)_6^{4-}$ in various water-alcohol mixtures have also been reported[255].

For $Mo(CN)_8^{4-}$, the quantum yield of hydrated electron formation[256] is practically constant (~ 0.30) upon irradiation in the 214–265 nm region and drops to ~0 at 365 nm, where $CN^-$ release occurs efficiently ($\Phi_{CN^-} \sim 0.8$)[257]. As for $Fe(CN)_6^{4-}$, it was assumed that electron ejection occurs from a CTTS excited state (~ 270 nm), which can be reached either directly or by internal conversion from the upper intramolecular CT state[256, 258]. Similar results have recently been reported for $Ru(CN)_6^{4-}$ and $W(CN)_8^{4-}$[259]. In the photooxidation of solvated metal ions in their normal oxidation states (*e.g.*, $Fe^{2+}_{aq}$), the solvated electron has never been observed, although the reaction can be described as an electron transfer from the cation to the solvent[2, 251]. Hydrated electrons, however, have been produced with a quantum yield of about unity upon irradiation at ~ 300 nm of the low oxidation state ions $Co^+$, $Ni^+$, $Zn^+$, and $Cd^+$, obtained by $e^-_{aq}$ reduction of the corresponding divalent ions[260]. Photooxidation of $Cr^{2+}_{aq}$[261] and $V^{2+}_{aq}$[262] seem also to involve $e^-_{aq}$ production.

Recently, Meisel *et al.*[177] have shown that hydrated electrons are produced with low efficiency upon light absorption by the lowest excited state of $Ru(bpy)_3^{2+}$, *i.e.*, in a biphotonic process.

Steady ultraviolet irradiation of benzoyl-ferrocene in hydroxylic solvents was reported to cause the formation of solvated electrons[263]. However, no evidence of solvated electrons has been found in flash photolysis and $N_2O$ scavenger experiments[264].

There is still some controversy as to the fate of the electron ejected from $Fe(CN)_6^{4-}$ and other anions. The assumption that the reactions of the "photochemical" hydrated electron are identical with those of the "radiolysis" hydrated electron has been questioned[265]. A careful examination of the decay kinetics of the photochemical hydrated electron has suggested a model in which the electron diffuses through a considerable volume of the medium before recombining with its radical co-product, during which time it may react with scavengers or electrons and radicals generated in other geminate pairs[266].

### 11.3. Production of Excited States in Reactions of Solvated Electrons

The formation of an excited state upon reduction of a transition metal complex by hydrated electrons in pulse radiolysis experiments was first suggested by Baxendale *et al.*[267]. In the reduction of Co(III) complexes (which have $(\pi_M t_{2g})^6$ electronic configuration) by $e^-_{aq}$, they thought that a "low-spin" Co(II) complex of $(\pi_M t_{2g})^6$ $(\sigma^*_M e_g)^1$ configuration could be formed. Such a complex would be an electronically excited state of the normal "high-spin" Co(II) complexes which have $(\pi_M t_{2g})^5(\sigma^*_M e_g)^2$ configuration. Baxendale's hypothesis was claimed to be confirmed by Waltz and Pearson[268], who observed a weak transient absorption at $\lambda > 400$ nm upon the reaction of $e^-_{aq}$ with $Co(bpy)_3^{3+}$. Baxendale and Fiti[269] made similar observations but concluded that the transient was the electron-adduct to the ligand, $Co^{III}(bpy)_2(bpy^-)^{2+}$. Hoffman and Simic[270] were able to discover a much more intense transient absorption at 300 nm for the same reaction and by some similarity with the spectrum of $bpy^-$ they also assigned the transient as a $bpy^-$ radical coordinated to the Co(III) center. The decay of the transient was described as an intramolecular transfer of the added electron to the metal center, the long lifetime of the transient (0.29 s) was taken as an indication of barriers to the process. In spectroscopic terms, such a transient would be a $(\pi_M t_{2g})^6(\sigma^*_M e_g)^0$ $(\pi_L)^1$ MLCT excited state of the $(\pi_M t_{2g})^5(\sigma^*_M e_g)^2(\pi^*_L)^0$ ground state $Co(bpy)_3^{2+}$ complex. Recent results, however, have shown that the observed transient is simply the ground state $Co(bpy)_3^{2+}$ complex which is labile and thus undergoes a rapid loss of a bpy ligand[271].

The formation of a coordinated reduced nitrobenzoato ligand radical (or, in spectroscopic terms, a MLCT excited state) has been found in the reaction of $e^-_{aq}$ with pentaaminocobalt(III) complexes containing a mono- or di-nitrobenzoato ligand[272]. Such a coordinated radical disappears *via* unimolecular kinetics, which represents ligand-to-metal intramolecular electron transfer with formation of $Co^{2+}$ (in spectroscopic terms, a radiationless deactivation of the MLCT excited state). The value of the electron-transfer rate constant ($10^2$–$10^5 s^{-1}$) is dependent upon the number and position of the nitro groups substituted on the aromatic ring.

The best proof of excited state formation is, of course, luminescence emission. As $Ru(bpy)_3^{2+}$ is known to exhibit luminescence from the lowest excited state irrespective of the excitation wavelength (Section 9.2), chemiluminescence should be observed in the one-electron reduction of $Ru(bpy)_3^{3+}$ if (spectroscopic) excited states are formed. This was found to be the case in the reduction of $Ru(bpy)_3^{3+}$ by the hydroxide ion[273] and by hydrazine[273, 274] in aqueous solution and by $Ru(bpy)_3^+$ [11, 275] in acetonitrile solution. Chemiluminescence was also observed by Martin *et al.*[78] and Matheson *et al.*[83] in the reduction of $Ru(bpy)_3^{3+}$ by $e_{aq}^-$. In all cases, the chemiluminescence spectrum was found to agree well with the photoluminescence of $Ru(bpy)_3^{2+}$, so that there is no doubt that these reactions lead, at least in part, to the lowest excited state of $Ru(bpy)_3^{2+}$, $*Ru(bpy)_3^{2+}$. For the reaction of $Ru(bpy)_3^{3+}$ with $e_{aq}^-$, the chemiluminescence quantum yield (0.015)[78] taken together with the photoluminescence quantum yield (0.042)[129] and the unit quantum yield of formation of the lowest excited state upon light excitation[132] indicates that ~ 36% of the reducing electrons end up in the $*Ru(bpy)_3^{2+}$ excited state [reaction (101)].

$$Ru(bpy)_3^{3+} + e_{aq}^- \longrightarrow *Ru(bpy)_3^{2+} \quad 38\%^{83)} \tag{101}$$

$$Ru(bpy)_3^{3+} + e_{aq}^- \longrightarrow {}^{\neq}Ru(bpy)_3^{2+} \quad > 55\%^{83)} \tag{102}$$

$$Ru(bpy)_3^{3+} + e_{aq}^- \longrightarrow Ru(bpy)_3^{2+} \quad < 7\%^{83)} \tag{103}$$

A recent investigation[83] substantially confirms this value and seems also to indicate that most of the remaining electrons yield another excited state [$^{\neq}Ru(bpy)_3^{2+}$, reaction (102)], which (i) decays to the ground state more slowly than does $*Ru(bpy)_3^{2+}$ and (ii) cannot be populated by light excitation. In our opinion, the existence of such an "isolated" excited state in a molecule like $Ru(bpy)_3^{2+}$ is unlikely and needs further confirmation. Reaction (103) has, of course, a higher driving force than reactions (101) and (102). Therefore, the predominant formation of excited states over ground state[83] is an example of Marcus "inverted" behavior (Section 6).

## 12. Other Systems

### 12.1. Metal Ions

For the photooxidation-reduction reactions of metal cations, it is often difficult to establish whether the reaction is outer or inner sphere, especially when the partner is a solvent molecule. Most of the available material has already been reviewed[2]. We will only take into consideration some recent papers which either are of general interest or clearly concern outer-sphere reactions.

*Vanadium.* $V_{aq}^{2+}$ in acidic solution is oxidized by UV light with concomitant $H_2$ evolution or reduction of $C_2H_2$, $C_2H_4$ and $N_2$[276]. The photocatalytic evolution of $H_2$ from water-alcohol solutions of $VCl_3$ has also been reported[262]. This study, which is related to the energy conversion problem, has shown that $V^{3+}$ is reduced ($\lambda = 280$ nm) in an alcohol solution with $\Phi = 0.23$, yielding $V^{2+}$ and aldehyde. The $V^{2+}$ formed is a strong reductant which under UV irradiation ($\lambda < 330$ nm) is able to undergo oxidation with formation of a solvated electron. Subsequent reaction of $e_s^-$ with $H^+$ causes the liberation of $H_2$ ($\Phi\ H_2 = 0.013$ at 280 nm). Normal hydrogen catalysts can be used for $V^{2+}$ oxidation, thus increasing the rate of $H_2$ evolution.

The process consists of the vanadium photocatalyzed decomposition of $C_2H_5OH$ into $CH_3CHO + H_2$, which is an endothermic reaction and thus leads to some conversion and storage of light energy into chemical energy. It should be noted, however, that the reaction requires 280-nm light and therefore it is not useful for solar energy utilization[103)]. Photoreduction of V(V) solutions with concomitant formation of $O_2$ has also been reported[277)].

*Manganese.* Photooxidation of $Mn^{2+}$ upon flash excitation in an aqueous solution has been reported[278)]. The redox interaction between $*Mn^{2+}$ and thionine[278)], in our opinion, is unlikely because $*Mn^{2+}$ is expected to be very short lived and the thionine concentration was very low.

*Iron.* Photoreduction of $Fe^{3+}$ in the presence of organic molecules in fluid solution takes place upon irradiation with UV light[279–283)]. Two types of reactions are thought to occur in these systems: (i) a charge-transfer transition from a water centered orbital to an iron centered orbital with the formation of OH radicals, which then react with organic substrate; and (ii) outer-sphere oxidation of the organic molecules by a charge-transfer excited state of the iron(III) species. The second type of reaction occurs with alcohols[281)], formic acid[281)] and ethylene glycol[283)]. Photooxidation of organic molecules by Fe(III) in rigid media at 77 K has been studied by the ESR technique[284)]. The results indicate that Fe(III) behaves more like Ce(IV) than $UO_2^{2+}$. The quantum yield of OH formation in the photoreduction of Fe(III) in water has recently been reported to be ~ 0.02 in the 436–297 nm range[285)]. Finally, the mechanism of oxidation of $Fe^{2+}$ in glassy aqueous solution at 77 K has been discussed[286)].

*Cobalt.* Electron transfer from $*Co^{2+}$ to thionine has recently been suggested to occur upon flash excitation[278)]. This process, however, is unlikely not only for the reasons mentioned before for $*Mn^{2+}$, but also because the lowest excited state of $*Co^{2+}$ (8.1 kK)[15)] is not expected to be a reductant strong enough to reduce thionine.

*Silver.* Silver perchlorate is photoreduced in aqueous and alcoholic solutions upon irradiation with 254 nm light[287)]. The mechanism is assumed to be electron transfer from a solvent molecule to a silver ion.

*Cerium.* Photooxidation of alcohols[288)] and carboxylic acids[289)] by Ce(IV) at 77 K proceeds *via* the formation of Ce(III) and organic radicals, which have been carefully studied by ESR technique. In aqueous nitric acid solution, $NO_3$ radicals are formed[290)]. Photooxidation of $Hg^+$ by Ce(IV) in an aqueous perchloric acid solution seems to involve the formation of an OH radical[291)].

*Europium.* The $Eu_{aq}^{2+}$ ion shows broad absorption bands at 250 and 320 nm which correspond to 4 f → 5d transitions having appreciable MLCT (CTTS) character[292)]. Irradiation at 254 nm causes photooxidation of $Eu^{2+}$ with concomitant formation of $H_2$[292–295)]. As the oxidation potential of $Eu^{2+}$ is 0.35 V, this photoreaction is thermodynamically allowed even in the dark and thus, contrary to a recent claim[295)], cannot be useful for light energy conversion (Section 8). There is some disagreement about the quantum yield of the photoreaction[292, 294, 295)], probably because different experimental conditions were used. The reaction mechanism seems to involve electron donation to a proton with formation of H atoms, which is followed by a complicated series of dark reactions[292, 293, 295)]. In 9 M aqueous $HClO_4$, irradiation leads to an $Eu^{3+}$ and a $ClO_3$ radical[296]. The europic ion, $Eu^{3+}_{aq}$, shows a broad absorption band in the near UV region, which has been assigned to a CT transition from water to the ion[297]. Irradiation in this spectral region causes the reduction of $Eu^{3+}$ to $Eu^{2+}$ [293, 297]. As $Eu^{2+}$ has an extinction coefficient much higher than $Eu^{3+}$, photooxidation of $Eu^{2+}$ becomes important after short irradiation periods[293, 297]. The photochemistry of the $Eu^{3+}/Eu^{2+}$ couple for energy conversion has been recently discussed[298]. The behavior of this system is somewhat similar to that of the $V^{3+}/V^{2+}$ system described above.

## 12.2. Miscellaneous

*Metallocenes.* Ferrocene[299, 300], ruthenocene[300, 301], and carbonyl-($\eta^5$-cyclopentadienyl)iron(I) tetramer[302] in halocarbon solvents give rise to charge transfer complexes whose excitation causes the formation of the corresponding complex cations. Excitation of ferrocene at 254 nm in ethanol or cyclohexane in the presence of $N_2O$ yields $N_2$ [303]. The most plausible mechanism is thought to be electron transfer from an excited state of the complex to $N_2O$. The same mechanism has also been proposed for the oxidation of excited carboxyalkylferrocenes by $N_2O$ in an aqueous solution[304].

*Cobalt.* Outer sphere oxidations of $Fe(CN)_6^{4-}$ by a MC excited state of $Co(en)_3^{3+}$ [305] and alcohol by a LMCT excited state of $Co(EDTA)^-$ [306] have been reported.

*Molybdenum.* Amine molybdates exhibit photochromic properties in the solid state[307–309] and in solution[309]. There is disagreement on whether photochromism is only exhibited by molybdates of secondary amines[308, 309]. UV irradiation generally causes a change in color from white to blue which is ascribable to the photoreduction of Mo(VI) to Mo(V) both in solid and in solution. The inhibiting effect of oxygen on the coloration is due to the rapid back oxidation of Mo(V) to Mo(VI).

Ultraviolet irradiation ($\lambda = 254$ nm) of $Mo_2(SO_4)_4^{4-}$ in 5 M $H_2SO_4$ produces $H_2$ and the one-electron oxidation product $Mo_2(SO_4)_4^{3-}$ [310]. The disappearance quantum yield is 0.17.

*Platinum.* Photolysis of some Pt(IV) complexes in aqueous solutions or in a frozen matrix results in the outer-sphere oxidation of anions and reduction of platinum to Pt(II)[311–314], probably *via* a Pt(III) intermediate.

*Acknowledgement.* We thank the National Research Council of Italy for its financial support.

# 13. References

1) Quinkert, G.: Angew. Chem., Int. Ed. Engl. *11*, 1072 (1972)

2) Balzani, V., Carassiti, V.: Photochemistry of coordination compounds. London: Academic Press 1970

3) Adamson, A. W., Fleischauer, P. D. (eds.): Concepts of Inorganic Photochemistry. New York: Wiley 1975
4) Balzani, V., Moggi, L., Manfrin, M. F., Bolletta, F., Laurence, G. S.: Coord. Chem. Rev. *15*, 321 (1975)
5) Koester, V. J.: Chem. Phys. Lett. *32*, 575 (1975)
6) Bock, C. R., Koerner von Gustorf, K.: Adv. Photochem. *10*, 221 (1977)
7) Meyer, T. J.: Acc. Chem. Res. *11*, 94 (1978)
8) Endicott, J. F.: Surv. Prog. Chem. *7*, 41 (1976)
9) Crosby, G. A.: Account Chem. Res. *8*, 231 (1975)
10) DeArmond, M. K.: Account Chem. Res. *7*, 309 (1974)
11) Tokel-Takvoryan, N. E., Hemingway, R. E., Bard, A. J.: J. Amer. Chem. Soc. *95*, 6582 (1973)
12) Corsby, G. A., Watts, R. J., Carstens, D. H. W.: Science *170*, 1195 (1970)
13) Demas, J. N., Harris, E. W., McBride, R. P.: J. Amer. Chem. Soc. *99*, 3547 (1977)
14) DeArmond, M. K., Hillis, J. E.: J. Chem. Phys. *54*, 2247 (1971)
15) Jørgensen, C. K.: Adv. Chem. Phys. *5*, 33 (1963)
16) Jørgensen, C. K.: Mol. Phys. *2*, 309 (1959)
17) Hipps, K. W., Crosby, G. A.: J. Amer. Chem. Soc. *97*, 7042 (1975)
18) Crosby, G. A.: Adv. Chem. Ser. *150*, 149 (1976)
19) Engel, P. S., Monroe, B. M.: Adv. Photochem. *8*, 245 (1971)
20) Lamola, A. A. in: Energy transfer and organic photochemistry. Lamola, A. A., Turro, N. J. (eds). New York: Wiley 1969, p. 17
21) Ermolaev, V. L., Sveshnikova, E. G., Shakverdov, T. A.: Russ. Chem. Rev. *44*, 26 (1975)
22) Ware, W. R., Lewis, C.: Mol. Photochem. *5*, 261 (1973)
23) Birks, J. B.: Photophysics of aromatic molecules. New York: Wiley 1970
24) Birks, J. B. (ed.): Organic molecular photophysics. New York: Wiley 1975, Vol. 2
25) Porter, G. B., Balzani, V., Moggi, L.: Adv. Photochem. *9*, 147 (1974)
26) Debye, P.: Trans. Electrochem. Soc. *82*, 265 (1942)
27) Eigen, M.: Z. Physik. Chem., Neue Folge *1*, 176 (1954)
28) Cotton, F. A., Wilkinson, G.: Advanced inorganic chemistry. New York: Wiley 1972, 3th edit.
29) Keeton, D. P., Basolo, F.: Inorg. Chim. Acta *6*, 33 (1972)
30) Maestri, M., Bolletta, F., Moggi, L., Balzani, V., Henry, M. S., Hoffman, M. Z.: J. Amer. Chem. Soc. in press
31) Sandrini, D., Gandolfi, M. T., Moggi, L., Balzani, V.: J. Amer. Chem. Soc. *99*, 4523 (1977)
32) Langford, C. H., Tong, J. P. K.: J. Chem. Soc., Chem. Commun. *1977*, 138
33) Wong, C. F. C., Kirk, A. D.: Can. J. Chem. *53*, 419 (1975)
34) Peterson, S. H., Demas, N. J.: J. Amer. Chem. Soc. *98*, 7880 (1976)
35) Giordano, P. J., Bock, C. R., Wrighton, M. S., Interrante, L. V., Williams, R. F. X.: J. Amer. Chem. Soc. *99*, 3187 (1977)
36) Vander Donckt, E. in: Elements de photochimie avancée. Courtot, P. (ed.). Paris: Hermann 1972, p. 79
37) Wagner, P. J.: Topics Curr. Chem. *66*, 1 (1976)
38) Grellmann, K. H., Watkins, A. R., Weller, A.: J. Phys. Chem. *76*, 469 (1972)
39) Labianca, D. A., Taylor, G. N., Hammond, G. S.: J. Am. Chem. Soc. *94*, 3679 (1972)
40) Parker, C. A.: Photoluminescence of solutions. Amsterdam: Elsevier 1968
41) Roy, J. K., Carroll, F. A., Whitten, D. G.: J. Am. Chem. Soc. *96*, 6349 (1974)
42) Stevens, B.: Adv. Photochem. *8*, 161 (1971)
43) Ware, W. R., Watt, D., Holmes, J. D.: J. Am. Chem. Soc. *96*, 7853 (1974)
44) Hino, T., Akazawa, H., Masuhara, H., Mataga, N.: J. Phys. Chem. *80*, 33 (1976)
45) Ballardini, R., Varani, G., Moggi, L., Balzani, V.: J. Am. Chem. Soc. *96*, 7123 (1974)
46) Ballardini, R., Varani, G., Moggi, L., Balzani, V.: J. Am. Chem. Soc. *99*, 6881 (1977)
47) Roy, J. K., Whitten, D. G.: J. Am. Chem. Soc. *93*, 7093 (1971)
48) Whitten, D. G., Meyer, T. J., Hopf, F. R., Ferguson, J. A., Brown, G.: Ann. N. Y. Acad. Sci. *206*, 516 (1973)

49) Adamson, A. W. in: Concepts of inorganic photochemistry. Adamson, A. W., Fleischauer, P. D. (eds.). New York: Wiley 1975, p. 413
50) Grabowski, Z. R., Grabowska, A.: Z. Physik. Chem. Neue Folge *101,* 197 (1976)
51) Grabowski, Z. R., Rubaszewska, W.: J. Chem. Soc., Faraday Trans. 1, *73,* 11 (1977)
52) Förster, Th.: Z. Electrochem. *54,* 531 (1950)
53) Bock, C. R., Meyer, T. J., Whitten, D. G.: J. Amer. Chem. Soc. *97,* 2909 (1975)
54) Ballardini, R., Varani, G., Indelli, M. T., Scandola, F., Balzani, V.: J. Amer. Chem. Soc., submitted
55) Memming, R.: Photochem. Photobiol. *16,* 325 (1972)
56) Gerischer, H.: Ber. Bunsenges. Phys. Chem. *77,* 771 (1973)
57) Maricle, D. L., Maurer, A. H.: Chem. Phys. Lett. *2,* 602 (1968)
58) Clark, W. D. K., Sutin, N.: J. Amer. Chem. Soc. *99,* 4676 (1977)
59) Zweig, A.: Adv. Photochem. *6,* 425 (1968)
60) Hercules, D. M. in: Techniques of chemistry. Weissberger, A., Rossiter, B. W. (eds.). New York: Wiley 1971, Vol. 1, Part IIB, p. 257
61) Taube, H.: Adv. Inorg. Chem. Radiochem. *1,* 1 (1959)
62) Halpern, J.: Quart. Rev. *15,* 217 (1961)
63) Sutin, N.: Acc. Chem. Res. *1,* 225 (1968)
64) Sutin, N.: A. Rev. Phys. Chem. *17,* 119 (1966)
65) Reynolds, W. L., Lumry, R. W.: Mechanisms of electron transfer. New York: The Ronald Press 1966
66) Basolo, F., Pearson, R. G.: Mechanisms of Inorganic Reactions. New York: Wiley 1967, p. 454
67) Sykes, A. G.: Adv. Inorg. Chem. Radiochem. *10,* 153 (1967)
68) Sutin, N. in: Inorganic biochemistry. Eichhorn, G. L. (ed.). Amsterdam: Elsevier 1973, Vol. 2, p. 611
69) Hush, N. S.: Prog. Inorg. Chem. *8,* 391 (1967)
70) Meyer, T. J.: Adv. Chem. Ser. *150,* 73 (1976)
71) Marcus, R. A.: A. Rev. Phys. Chem. *15,* 155 (1964)
72) Marcus, R. A.: J. Chem. Phys. *24,* 966 (1956): *43,* 679 (1965)
73) Marcus, R. A.: J. Phys. Chem. *72,* 891 (1968)
74) Marcus, R. A.: J. Chem. Phys. *43,* 2654 (1965); *52,* 2803 (1970)
75) Marcus, R. A.: Discuss. Faraday Soc. *29,* 21 (1960)
76) Haim, A., Sutin, N.: Inorg. Chem. *15,* 476 (1976)
77) Sutin, N.: A. Rev. Nucl. Sci., *12,* 285 (1962)
78) Martin, J. E., Hart, E. J., Adamson, A. W., Gafney, H., Halpern, J.: J. Am. Chem. Soc. *94,* 9238 (1972)
79) Chou, M., Creutz, C., Sutin, N.: J. Am. Chem. Soc. *99,* 5615 (1977)
80) Mentasti, E., Pelizzetti, E., Baiocchi, C.: J. Chem. Soc., Dalton Trans. *1977,* 132, and references therein
81) White, E. H., Miano, J. D., Watkins, C. J., Breaux, J. E.: Angew. Chem. Int. Ed. Engl. *13,* 229 (1974)
82) Herculés, D. M.: Acc. Chem. Res. *2,* 301 (1969)
83) Jonah, C. D., Matheson, M. S., Meisel, D.: J. Am. Chem. Soc. *100,* 1449 (1978)
84) Henglein, A., Grätzel, M. in: Solar power and fuels. Bolton, R. (ed.). New York: Academic Press 1977, p. 53
85) Rehm, D., Weller, A.: Ber. Bunsenges. Phys. Chem. *73,* 834 (1969)
86) Rehm, D., Weller, A.: Isr. J. Chem. *8,* 259 (1970)
87) Creutz, C., Sutin, N.: J. Am. Chem. Soc. *99,* 241 (1977)
88) Efrina, S., Bixon, M.: Chem. Phys. Lett. *25,* 34 (1974); Chem. Phys. *13,* 447 (1976)
89) Marcus, R. A., Sutin, N.: Inorg. Chem. *14,* 213 (1975)
90) Van Duyne, R P., Fisher, S. F.: Chem. Phys. *5,* 183 (1974)
91) Schmickler, W.: J. Chem. Soc., Faraday Trans. 2, *72,* 307 (1976)
92) Dogonadze, P. R., Kuznetsov, A. M.: Z. Physik. Chem., Neue Folge *100,* 1 (1976)
93) Vogelmann, E., Schreiner, S., Rauscher, W., Kramer, H. E. A.: Z. Phys. Chem., Neue Folge *101,* 321 (1976)

94) Scheerer, R., Grätzel, M.: J. Am. Chem. Soc. *99,* 865 (1977)
95) Schulten, K., Staerk, H., Weller, A., Werner, H. J., Nickel, B.: Z. Phys. Chem. *101,* 371 (1976)
96) Eigen, M., Kruse, W., Maas, G., De Maeyer, L.: Prog. React. Kinet. *2,* 285 (1964)
97) Jarnagin, R. C.: Acc. Chem. Res. *4,* 420 (1971)
98) Meyer, T. J.: Isr. J. Chem., *15,* 200 (1977)
99) Young, R. C., Meyer, T. J., Whitten, D. G.: J. Am. Chem. Soc. *97,* 4781 (1975)
100) Anderson, C. P., Salomon, D. J., Meyer, T. J., Young, R. C.: J. Am. Chem. Soc. *99,* 1980 (1977)
101) Young, R. C., Keene, F. R., Meyer, T. J.: J. Am. Chem. Soc. *99,* 2468 (1977)
102) Nagle, J., Young, R. C., Meyer, T. J.: Inorg. Chem. *16,* 3366 (1977)
103) Balzani, V., Moggi, L., Manfrin, M. F., Bolletta, F., Gleria, M.: Science, *189,* 852 (1975)
104) Archer, M. D.: Photochemistry *6,* 739 (1975)
105) Archer, M. D.: Photochemistry *7,* 561 (1976)
106) Bolton, J. R., (ed.): Solar power and fuels. New York, Academic Press 1977
107) Albery, J. W., Archer, M. D.: Nature *270,* 399 (1977)
108) Archer, M. D.: J. Appl. Electrochem. *5,* 17 (1975)
109) Lichtin, N. N. in: Solar power and fuels. Bolton, J. R., (ed.). New York: Academic Press, 1977, p. 119
110) Gomer, R.: Electrochim. Acta *20,* 13 (1975)
111) Moggi, L. in: Solar power and fuels. Bolton, J. R. (ed.). New York: Academic Press 1977, p. 147
112) Ti Tien, H., Karvaly, B. in: Solar power and fuels. Bolton, J. R. (ed.). New York: Academic Press 1977, p. 167
113) McWhinnie, W. R., Miller, J. D.: Adv. Inorg. Chem. *12,* 135 (1969)
114) Gafney, D. H., Adamson, A. W.: J. Am. Chem. Soc. *94,* 8238 (1972)
115) Natarajan, P., Endicott, J. F.: J. Phys. Chem. *77,* 971 (1973)
116) Natarajan, P., Endicott, J. F.: J. Phys. Chem. *77,* 1823 (1973)
117) Natarajan, P., Endicott, J. F.: J. Am. Chem. Soc. *94,* 3635 (1972)
118) Natarajan, P., Endicott, J. F.: J. Am. Chem. Soc. *95,* 2470 (1973)
119) Natarajan, P., Endicott, J. F.: J. Phys. Chem. *77,* 2049 (1973)
120) Breck, A. K., Wan, J. K. S.: Int. J. Radiat. Phys. Chem. *5,* 517 (1973)
121) Bock, C. R., Meyer, T. J., Whitten, D. G.: J. Am. Chem. Soc. *96,* 4710 (1974)
122) Navon, G., Sutin, N.: Inorg. Chem. *13,* 2159 (1974)
123) Laurence, G. S., Balzani, V.: Inorg. Chem. *13,* 2976 (1974)
124) Saji, T., Aoyagui, S.: J. Electroanal. Chem. Interfacial Electrochem. *58,* 401 (1975)
125) Saji, T., Aoyagui, S.: J. Electroanal. Chem. Interfacial Electrochem. *60,* 1 (1975)
126) Roffia, S., Ciano, M.: J. Electroanal. Chem. *77,* 349 (1977) and unpublished results
127) DeArmond, M. K., Hanck, K. W.: private communication
128) Maestri, M., Bolletta, F., Serpone, N., Moggi, L., Balzani, V.: Inorg. Chem. *15,* 2048 (1976), and refs. therein
129) Van Houten, J., Watts, R. J.: J. Am. Chem. Soc. *98,* 4853 (1976)
130) Hoggard, P. E., Porter, G. B.: J. Am. Chem. Soc. *100,* 1457 (1978)
131) Gleria, M., Minto, F., Beggiato, G., Bortolus, P.: J. Chem. Soc., Chem. Commun. *1978,* 285
132) Bolletta, F., Maestri, M., Balzani, V.: J. Phys. Chem. *80,* 2499 (1976)
133) Demas, N. J., Crosby, G. A.: J. Am. Chem. Soc. *93,* 2841 (1971)
134) Bensasson, R., Salet, C., Balzani, V.: J. Am. Chem. Soc. *98,* 3722 (1976) and unpublished results
135) Watts, R. J., Crosby, G. A., Sansregret, J. L.: Inorg. Chem. *11,* 1474 (1972)
136) Watts, R. J., Crosby, G. A.: J. Am. Chem. Soc. *93,* 3184 (1971)
137) König, E., Herzog, S.: J. Inorg. Nucl. Chem. *32,* 585 (1970)
138) Kane-Maguire, N. A. P., Conway, J., Langford, C. H.: J. Chem. Soc. Chem. Commun. *1974,* 801
139) Maestri, M., Bolletta, F., Moggi, L., Balzani, V., Henry, M. S., Hoffmann, M. Z.: J. Chem. Soc. Chem. Commun. *1977,* 492

140) Sutin, N., Creutz, C.: Adv. Chem. Ser., in press
141) Serpone, N., Bolletta, F., *et al.:* work in progress
142) Lytle, F. E., Hercules, D. M.: J. Am. Chem. Soc. *91,* 253 (1969)
143) Klassen, D. M., Crosby, G. A.: J. Chem. Phys. *48,* 1853 (1968)
144) Van Houten, J., Watts, R. J.: J. Am. Chem. Soc. *97,* 3843 (1975)
145) Allsopp, S. R., Cox, A., Jenkins, S. H., Kemp. T. J., Tunstall, S. M.: Chem. Phys. Lett. *43,* 135 (1976)
146) Harrigan, R. W., Crosby, G. A.: J. Chem. Phys. *59,* 3468 (1973)
147) Harrigan, R. W., Hager, G. D., Crosby, G. A.: Chem. Phys. Lett. *21,* 487 (1973)
148) Fujita, I., Kobayashi, H.: Inorg. Chem. *12,* 2758 (1973)
149) Král, M.: Theor. Chim. Acta *43,* 273 (1977)
150) Král, M.: Naturwissenschaften *64,* 385 (1977)
151) Watts, R. J., Harrington, J. S., Van Houten, J.: Adv. Chem. Ser., in press
152) Mahon, C., Reynolds, W. L.: Inorg. Chem. *6,* 1927 (1967)
153) Crosby, G. A., Elfrin, W. H. Jr.: J. Phys. Chem. *80,* 2206 (1976)
154) Lin, C-T., Sutin, N.: J. Phys. Chem. *80,* 97 (1976)
155) Lin, C-T., Sutin, N.: J. Am. Chem. Soc. *97,* 3543 (1975)
156) Lin, C-T., Bottcher, W., Chou, M., Creutz, C., Sutin, N.: J. Am. Chem. Soc. *98,* 6536 (1976)
157) Balzani, V., Moggi, L., Bolletta, F., Manfrin, M. F.: Adv. Chem. Ser. *150,* 160 (1976)
158) Ballardini. R., Varani, G., Scandola, F., Balzani, V.: J. Am. Chem. Soc. *98,* 7432 (1976)
159) Balzani, V., Bolletta, F., Moggi, L.: Spectros. Lett., in press
160) Juris, A., Manfrin, F., Maestri, M., Serpone, N.: Inorg. Chem., in press
161) Demas, J. N., Addington, J. W.: J. Am. Chem. Soc. *98,* 5800 (1976)
162) Demas, J. N., Addington, J. W., Peterson, S. H., Harris, E. W.: J. Phys. Chem. *81,* 1039 (1977)
163) Creutz, C.: Inorg. Chem., submitted
164) Toma, H. E., Creutz, C.: Inorg. Chem., *16,* 545 (1977)
165) Juris, A., Gandolfi, M. T., Manfrin, M. F., Balzani, V.: J. Am. Chem. Soc. *98,* 1047 (1976)
166) Creutz, C., Sutin, N.: Inorg. Chem. *15,* 496 (1976)
167) Gutierrez, A. R., Meyer, T. J., Whitten, D. G.: Mol. Photochem. *7,* 349 (1976)
168) Demas, J. N., Harris, E. W., Flynn, C. M., Diemente, D.: J. Am. Chem. Soc. *97,* 3838 (1975)
169) Winterle, J. S., Kliger, D. S., Hammond, G. S.: J. Am. Chem. Soc. *98,* 3719 (1976)
170) Phillips, J., Langford, C. H., Konigstein, J. A.: J. Chem. Soc., Chem. Commun. *1977,* 425
171) Meisel, D., Matheson, M.: J. Am. Chem. Soc. *99,* 6577 (1977)
172) Chandrasekaran, K., Natarajan, P.: J. Chem. Soc., Chem. Commun. *1977,* 774
173) Young, R. C., Meyer, T. J., Whitten, D. G.: J. Am. Chem. Soc. *98,* 286 (1976)
174) Mc Millin, D. R., Bucker, M. T., Ahn, B. T.: Inorg. Chem. *16,* 943 (1977)
175) Maestri, M., Grätzel, M.: Ber. Bunsenges. Phys. Chem. *81,* 504 (1977)
176) Langford, C. H., Vuik, C., Kane-Maguire, N. A. P.: Inorg. Nucl. Chem. Lett. *11,* 377 (1975)
177) Meisel, D., Matheson, M. S., Mulac, W. A., Rabani, J.: J. Phys. Chem. *81,* 1449 (1977)
178) Wehry, E. L., Ward, A. R.: Inorg. Chem. *10,* 2660 (1971)
179) Demas, J. N., Adamson, A. W.: J. Am. Chem. Soc. *95,* 5159 (1973)
180) Lottaz, A., Zelewsky, A.: Chimia *31,* 180 (1977)
181) Bolletta, F., Maestri, M., Moggi, L., Balzani, V.: J. Chem. Soc., Chem. Commun. *1975,* 901
182) Lachish, U., Ottolenghi, M., Rabani, J.: J. Am. Chem. Soc. *99,* 8062 (1977)
183) Creutz, C., Sutin, N.: J. Am. Chem. Soc. *98,* 6384 (1976)
184) Delaive, P. J., Lee, J. T., Sprinschnik, H. W., Abruna, H., Meyer, T. J., Whitten, D. G.: J. Am. Chem. Soc. *99,* 7094 (1977)
185) Delaive, P. J., Lee, J. T., Abruna, H., Sprintschnik, H. W., Meyer, T. J., Whitten, D. G.: Adv. Chem. Ser., in press
186) Clark, W. D. K., Eckert, J. A.: Sol. Energy *17,* 147 (1975)
187) Wildes, P. D., Hobart, D. R., Lichtin, N. N., Hall, D. E., Eckert, A. J.: Sol. Energy *19,* 567 (1977)
188) Shigehara, K., Tsuchida, E.: J. Phys. Chem. *81,* 1883 (1977)
189) Creutz, C., Sutin, N.: Proc. Natl. Acad. Sci. U.S.A. *72,* 2858 (1975)

190) Sprintschnik, G., Sprintschnik, H. W., Kirsch, P. P., Whitten, D. G.: J. Am. Chem. Soc. *98,* 2337 (1976)
191) Whitten, D. G.: Pure Appl. Chem. *49,* 379 (1977)
192) Valenty, S. J., Gaines, G. L. Jr.: J. Am. Chem. Soc. *99,* 1285 (1977)
193) Sprintschnik, G., Sprintschnik, H. W., Kirsch, P. P., Whitten, D. G.: J. Am. Chem. Soc. *99,* 4947 (1977)
194) Harriman, A.: J. Chem. Soc., Chem. Commun. *1977,* 777
195) Gleria, M., Memming, R.: Z. Phys. Chem. Neue Folge *98,* 303 (1975)
196) Clark, D. K., Sutin, N.: J. Am. Chem. Soc. *99,* 4676 (1977)
197) Maestri, M., *et al.:* work in progress
198) Kayser, R. H., Young, R. H.: Photochem. Photobiol. *24,* 395 (1976)
199) Kayser, R. H., Young, R. H.: Photochem. Photobiol. *24,* 403 (1976)
200) Endicott, J. F., Ferraudi, G. J.: J. Am. Chem. Soc. *99,* 5812 (1977)
201) Rabinowitch, E., Belford, R. L.: Spectroscopy of uranil compounds. Oxford: Pergamon Press 1964
202) Burrows, H. D., Kemp, T. J.: Chem. Soc. Rev. *3,* 139 (1974)
203) McGlynn, S. P., Smith, J. K.: J. Mol. Spectros. *6,* 164 (1961)
204) Bell, J. T., Biggers, R. E.: J. Mol. Spectros. *18,* 247 (1965)
205) Bell, J. T.: J. Inorg. Nucl. Chem. *31,* 703 (1969)
206) Brint, P., McCaffery, A. J.: Mol. Phys. *25,* 311 (1973)
207) Volod'ko, L. V., Komyak, A. I., Sleptsov, L. E.: Opt. Spectrosc. *23,* 397 (1967)
208) Gorller-Walrand, C., Vanquickenborne, L. G.: J. Phys. Chem. *54,* 4178 (1971); *idem 57,* 1436 (1972)
209) Jorgensen, C. K., Reisfeld, R.: Chem. Phys. Lett. *35,* 441 (1975)
210) Jorgensen, C. K.: Naturwissenschaften *64,* 37 (1977)
211) Jorgensen, C. K.: Prog. Inorg. Chem. *12,* 101 (1970)
212) Benson, P., Cox, A., Kemp, T. J., Sultana, Q.: Chem. Phys. Lett. *35,* 195 (1975)
213) Yokoyama, Y., Moriyasu, M., Ikeda, S.: J. Inorg. Nucl. Chem. *38,* 1329 (1976)
214) Burrows, H. D., Formosinho, S. J., Da Graça Miguel, M., Pinto Coelho, F.: J. Chem. Soc. Faraday Trans. 1, *72,* 163 (1976)
215) Hill, R. J., Kemp, T. J., Allen, D. M., Cox, A.: J. Chem. Soc. Faraday Trans. 1, *70,* 847 (1974)
216) Sergeeva, G. I., Chibisov, A. K., Levshin, L. V., Karyakin, A. V.: J. Chem. Soc. Chem. Commun. *1974,* 159
217) Sergeeva, G. I., Chibisov, A. K., Karyakin, A. V., Levshin, L. V., Nemodruk, A. A., Myasoedov, B. F.: Kim. Vys. Energ. *8,* 38 (1974)
218) Allen, D. M., Burrows, H. D., Cox, A., Hill, R. J., Kemp, T. J., Stone, T. J.: J. Chem. Soc. Chem. Commun. *1973,* 59
219) Handbook of Chemistry and Physics. Cleveland: Chemical Rubber Co. 1975–1976, 56th edit.
220) Rygalov, L. N., Chibisov, A. K., Karyakin, A. V., Bezrogova, E. V., Myasoedov, B. F., Nemodruk, A. A.: Teor. Eksp. Khim. *8,* 484 (1972)
221) Feve, M.: C. R. Acad. Sci., Ser. C, *279,* 721 (1974)
222) Sergeeva, G. I., Chibisov, A. K., Levshin, L. V., Karyakin, A. V., Myasoedov, B. F., Nemodruk, A. A.: Khim. Vys. Energ. *8,* 358 (1974)
223) Kato, Y., Fukutomi, H.: J. Inorg. Nucl. Chem. *38,* 1323 (1976)
224) Markantonatos, M. D.: Inorg. Chim. Acta, *24,* L37 (1977)
225) Sakuraba, S., Matsushima, R.: Bull. Chem. Soc. Jpn. *43,* 2359 (1970)
226) Sakuraba, S., Matsushima, R.: Bull. Chem. Soc. Jpn. *44,* 2915 (1971)
227) Matsushima, R.: J. Am. Chem. Soc. *94,* 6010 (1972)
228) Greatorex, D., Richard, J. H., Kemp, T. J., Stone, T. J.: J. Chem. Soc. Faraday Trans. 1, *70,* 216 (1974)
229) Ledwith, A., Russel, P. J., Sutcliffe, L. H.: Proc. R. Soc. London, Ser. A *332,* 151 (1973)
230) Greatorex, D., Hill, R. J., Kemp, T. J., Stone, T. J.: J. Chem. Soc., Faraday Trans. 1, *68,* 2059 (1972)
231) Sakuraba, S., Matsushima, R.: Bull. Chem. Soc. Jpn. *43,* 1950 (1970); idem, *44,* 1278 (1971)

232) Sakuraba, S., Matsushima, R.: Chem. Lett. *1972*, 911
233) Matsushima, R., Mori, K., Suzuki, M.: Bull. Chem. Soc. Jpn. *49*, 38 (1976)
234) Burrows, H. D., Formosinho, S. J.: J. Chem. Soc., Faraday Trans. 1, *73*, 201 (1977)
235) Ahmed, M., Cox, A., Kemp, T. J., Sultana, Q.: J. Chem. Soc., Perkin, Trans. 2, *1975*, 1867
236) Yokoyama, Y., Moriyasu, M., Ikeda, S.: J. Inorg. Nucl. Chem. *36*, 385 (1974)
237) Matsushima, R., Fujimori, H., Sakuraba, S.: J. Chem. Soc., Faraday Trans. 1, *70*, 1702 (1974)
238) Markantonatos, M. D.: Inorg. Chim. Acta *24*, L 53 (1977)
239) Markantonatos, M. D.: Inorg. Chim. Acta *25*, L 87 (1977)
240) Markantonatos, M. D.: Inorg. Chim. Acta *25*, L 101 (1977)
241) Markantonatos, M. D.: Inorg. Chim. Acta *26*, 41 (1978)
242) Endicott, J. F. in: Concepts in inorganic photochemistry. Adamson, A. W., Fleischauer, P. D. (eds.). New York: Wiley 1975, p. 81
243) Berdinikov, B. M., Bazhim, N. M.: Zh. Fiz. Khim. *44*, 712 (1970)
244) Burrows, H. D., Pedrosa de Jesus, J. D.: J. Photochem. *5*, 265 (1976)
245) Stasicka, Z., Marchaj, A.: Coord. Chem. Rev. *23*, 131 (1977)
246) Hart, E. J., Anbar, M.: The hydrated electron. New York: Wiley 1970
247) Eiben, K.: Angew. Chem. Int. Ed. Engl. *9*, 619 (1970)
248) Matheson, M. S., Mulac, W. A., Rabani, J.: J. Phys. Chem. *67*, 2613 (1963)
249) Crawford, M., Rumfeldt, R.: Can. J. Chem. *49*, 3336 (1971)
250) Shirom, M., Stein, G.: J. Chem. Phys. *55*, 3372 (1971)
251) Fox, M. in: Concepts in inorganic photochemistry. Adamson, A. W., Fleischauer, P. D. (eds.). New York: Wiley 1975, p. 333
252) Stein, G.: Isr. J. Chem. *8*, 691 (1970)
253) Shirom, M., Weiss, M.: J. Chem. Phys. *56*, 3170 (1972)
254) Moam, J., Shirom, M., Weiss, M.: Isr. J. Chem. *11*, 21 (1973); C. A. *79*, 99154 (1973)
255) Shirom, M., Tomkiewicz, M.: J. Chem. Phys. *56*, 2731 (1972)
256) Shirom, M., Siderer, Y.: J. Chem. Phys. *57*, 1013 (1972)
257) Balzani, V., Manfrin, M. F., Moggi, L.: Inorg. Chem. *8*, 47 (1969)
258) Shirom, M., Siderer, Y.: J. Chem. Phys. *58*, 1250 (1973)
259) Kalisky, O., Shirom, M.: J. Photochem. *7*, 215 (1977)
260) Basco, N., Vidyarthi, S. K., Walker, D. C.: Can. J. Chem. *52*, 343 (1974)
261) Hartmann, H., Müller, J., Kelm, H.: Naturwissenschaften *60*, 256 (1973)
262) Koryakin, B. V., Dzabiev, T. S., Shilov, A. E.: Dokl. Phys. Chem. *229*, 614 (1976)
263) Traverso, O., Rossi, R., Sostero, S., Carassiti, V.: Mol. Photochem. *5*, 457 (1973)
264) Heaney, E. K., Logan, S. R., Powell, J. A.: J. Chem. Soc. Faraday Trans. 1, *73*, 699 (1977)
265) Bryant, F. D., Santus, R., Grossweiner, L. I.: J. Phys. Chem. *79*, 2711 (1975)
266) Grossweiner, L. I., Baugher, J. F.: J. Phys. Chem. *81*, 93 (1977)
267) Baxendale, J. H., Fielden, E. M., Keene, J. P.: Proc. R. Soc. London, Ser. A *286*, 320 (1965)
268) Waltz, W. L., Pearson, R. G.: J. Phys. Chem. *73*,, 1941 (1969)
269) Baxendale, J. H., Fiti, M.: J. Chem. Soc., Dalton Trans. *1972*, 1995
270) Hoffmann, M. Z., Simic, M.: J. Chem. Soc., Chem. Commun. *1973*, 640
271) Mulazzani, Q. G., Hoffmann, M. Z., Simic, M. G.: private communication
272) Simic, M. G., Hoffmann, M. Z. Brezmak, N. V.: J. Am. Chem. Soc. *99*, 2166 (1977)
273) Hercules, D. M., Lytle, F. E.: J. Am. Chem. Soc. *88*, 4745 (1966)
274) Lytle, F. E., Hercules, D. M.: Photochem. Photobiol. *13*, 123 (1971)
275) Tokel, N. E., Bard, A. J.: J. Am. Chem. Soc. *94*, 2862 (1972)
276) Zones, S. I., Vickrey, T. M., Palmer, J. G., Schrauzer, G. N.: J. Am. Chem. Soc. *98*, 7289 (1976)
277) Jeliazkowa, B. G., Nakamura, S., Fukutomi, H.: Bull. Chem. Soc. Jpn. *48*, 347 (1975)
278) Ferreira, M. I. C., Harriman, A.: J. Chem. Soc., Faraday Trans. 1, *73*, 1085 (1977)
279) Carey, J. H. Langford, C. H.: Inorg. Nucl. Chem. Lett. *10*, 591 (1974)
280) Langford, C. H., Carey, J. H.: Can. J. Chem. *53*, 2430 (1975)
281) Carey, J. H., Langford, C. H.: Can. J. Chem. *53*, 2436 (1975)
282) Carey, J. H., Oliver, B. G.: Can. J. Chem. *55*, 1207 (1977)
283) Carey, J. H., Cosgrove, E. G., Oliver, B. G.: Can. J. Chem. *55*, 625 (1977)

284) Cox, A., Kemp, T. J.: J. Chem. Soc., Faraday Trans. 1, *71,* 2490 (1975)
285) Balakrishnan, N. J. I., Reddy, M. P.: J. Phys. Chem. *81,* 17 (1977)
286) Korolev, V. V., Bazhin, N. M.: Khim. Vys. Energ. *8,* 506 (1974)
287) Hada, H., Yonezawa, Y., Yoshida, A., Kurakake, A.: J. Phys. Chem. *80,* 2728 (1976)
288) Greatorex, D., Kemp, T. J.: Trans. Faraday Soc. *67,* 56 (1971)
289) Greatorex, D., Kemp, T. J.: Trans. Faraday Soc. *67,* 1576 (1971)
290) Glass, R. W., Martin, T. W.: J. Am. Chem. Soc. *92,* 5084 (1970)
291) Nuyt, A., Zador, M.: Can. J. Chem. *50,* 2413 (1972)
292) Davis, D. D., Stevenson, K. L., King, G. K.: Inorg. Chem. *16,* 670 (1977)
293) Haas, Y., Stein, G., Tenne, R.: Isr. J. Chem. *10,* 529 (1972)
294) Bulgakov, R. G., Kazakov, V. P., Korobeinikova, V. N.: Opt. Spectrosc. *35,* 497 (1973)
295) Ryason, P. R.: Sol. Energy *19,* 445 (1977)
296) Kazakov, V. P., Bulgakov, R. G., Konoplya, M. M.: Khim. Vys. Energ. *10,* 181 (1976); C. A. *85,* 114643 (1976)
297) Haas, Y., Stein, G., Tomkiewicz, M.: J. Phys. Chem. *74,* 2558 (1970)
298) Stein, G.: Isr. J. Chem. *14,* 213 (1975)
299) Traverso, O., Scandola, F.: Inorg. Chim. Acta *4,* 493 (1970)
300) Barrell, P., Henderson, E.: J. Chem. Soc., Dalton Trans. *1975,* 432
301) Traverso, O., Sostero, S., Mazzocchin, G. A.: Inorg. Chim. Acta *11,* 237 (1974)
302) Bock, C. R., Wrighton, M. S.: Inorg. Chem. *16,* 1309 (1977)
303) Powell, J. A., Logan, S. R.: J. Photochem. *3,* 189 (1974)
304) Heaney, E. K., Logan, S. R.: J. Chem. Soc., Perkin Trans. 2, *1977,* 1353
305) Kane-Maguire, N. A. P., Langford, C. H.: J. Chem. Soc., Chem. Commun. *1973,* 351
306) Langford, C. H.: Can. J. Chem. *55,* 3132 (1977)
307) Yamase, T., Hayashi, H., Ikawa, T.: Chem. Lett. *1974,* 1055
308) Arnaud-Neu, F., Schwing-Weill, M. J.: J. Less-Common Met. *36,* 71 (1974), and references therein
309) Yamase, T., Ikawa, T.: Bull. Chem. Soc. Jpn. *50,* 746 (1977), and references therein
310) Erwin, D. K., Geoffroy, G. L., Gray, H. B., Hammond, G. S., Solomon, E. I., Trogler, W. C., Zagars, A. A.: J. Am. Chem. Soc. *99,* 3620 (1977)
311) Shagisultanova, G. A., Karaban, A. A., Orisheva, R. M., Gorbunova, S. P.: Prevrashch. Kompleks. Soedin. Deistviem. Sveta, Radiats. Temp. *1973,* 45; C. A. *81,* 31771 (1974)
312) Shagisultanova, G. A., Karaban, A. A., Orisheva, R. M., Gorbunova, S. P.: Khim. Vys. Energ. *7,* 56 (1973); C. A. *80,* 8968 (1974)
313) Shagisultanova, G. A., Karaban, A. A., Poznyak, A. L.: Khim. Vys. Energ. *5,* 368 (1971); C. A. *75,* 103604 (1971)
314) Shagisultanova, G. A., Karaban, A. A., Gorbunova, S. P.: Fotokhim. Radiats-Khim. Protsessy Vod. Rastvorakh Tverd. Telakh *1970,* 179; C. A. *75,* 135770 (1971)

Received February 21, 1978

# Photochemical Reactivity of Keto Imino Ethers

## Type I. Rearrangement and (2 + 2)-Photocycloaddition to the Carbon-Nitrogen Double Bond

**Tad H. Koch, David R. Anderson, John M. Burns, Geoffrey C. Crockett, Kent A. Howard, Joseph S. Keute, Ronald M. Rodehorst, and Richard J. Sluski**

Department of Chemistry, University of Colorado, Boulder, Colorado 80309, U.S.A.

**Table of Contents**

## I. Introduction

(2 + 2)-Photocycloaddition of olefins to carbon-carbon[1)] and carbon-oxygen double bonds[2)] are well-documented reactions in the chemical literature often employed in synthetic schemes[3)]. Analogous cycloadditions to carbon-nitrogen double bonds, however, are rare. This review describes our attempts and the attempts of others to define some structural and electronic prerequisites necessary to achieve the (2 + 2)-photocycloaddition reaction to carbon-nitrogen double bonds. During our pursuit of this elusive reaction, we discovered a facile Type I rearrangement reaction which will also be described in some detail.

Prior, to our work on (2 + 2)-photocycloaddition reactions to carbon-nitrogen double bonds, there were only three relevant reports, two of which have subsequently been shown to be in error. Irradiation of N-(4-dimethylaminobenzylidene)aniline (*1*) was reported to give *cis*-4,4′-bis-(dimethylamino)stilbene (*2*) and azobenzene (*3*) *via* an unstable 1,2-diazetidine intermediate (*4*)[4)]. Ohta and Tokumaru, however, have recently shown that this mode of reactivity is not reproducible[5)]. The only

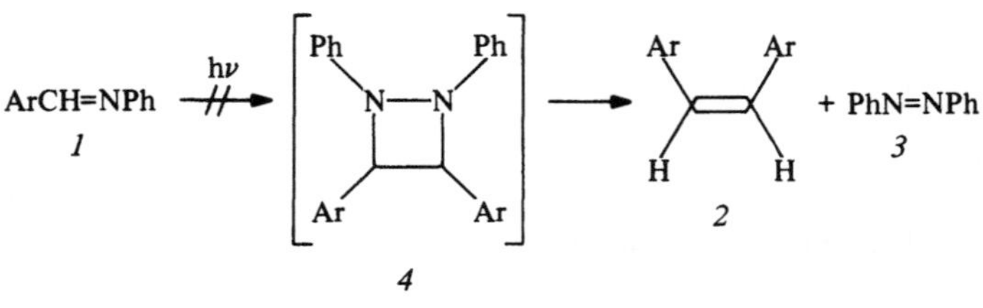


Ar≡p–$Me_2NC_6H_4$

products they observed from irradiation of *1* under similar conditions were 4-dimethylaminobenzaldehyde and aniline. A 1,3-diazetidine (*5*) was reported to be a stable photoproduct from the benzophenone sensitized irradiation of benzaldehyde N-cyclohexylimine (*6*) in ethanol or 2-propanol solvent[6)]. Padwa and co-workers later established the structure of the photoproduct as N,N′-dicyclohexyl-1,2-diamino-1,2-diphenylethane (*7*)[7)]. Tsuge and co-workers described the cycloaddition of 2,5-

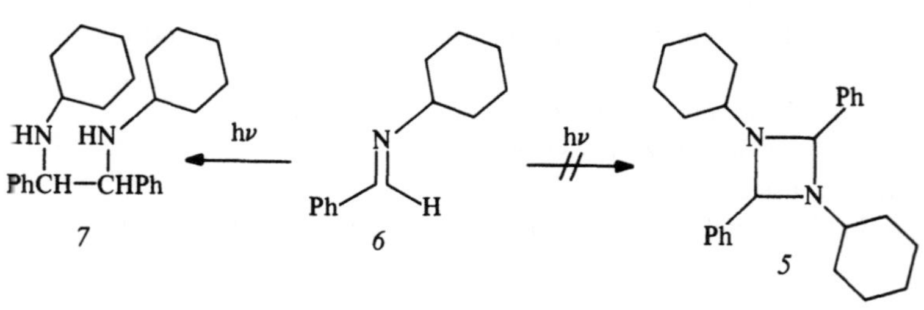


diphenyl-1,3,4-oxadiazole (*8*) to indene (*9*) in the presence of iodine to form the (2 + 2)-cycloadduct *10* and the 2:1 type adduct *11*, both in poor yield[8)]. In the absence of iodine only the 2:1 type adduct was observed. The reaction mechanism for formation of *10* and the role of iodine have not been clearly established[9)].

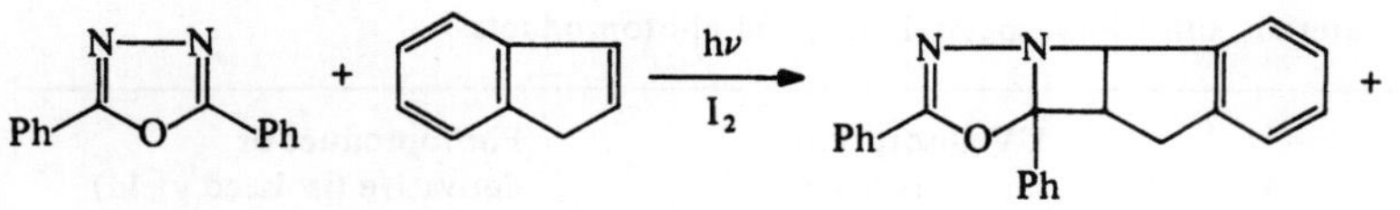

8 9 10

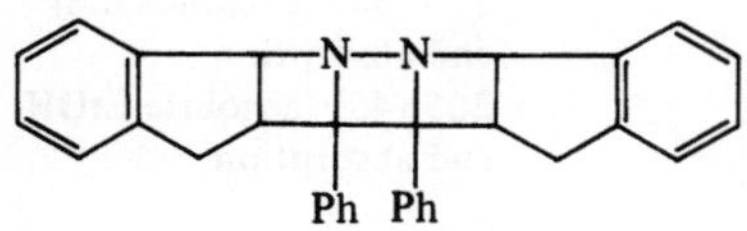

11

Our search for the (2 + 2)-photocycloaddition reaction to the carbon-nitrogen double bond has concentrated on the photo-chemical reactivity of a series of nitrogen heterocycles which structurally resemble cyclic α,β-unsaturated ketones. Specifically, we have studied cyclic molecules which possess carbonyl and imino ether functional groups in conjungation. This approach was selected because cyclic α,β-unsaturated ketones undergo facile (2 + 2)-photocycloaddition reactions[10–12]. The possibility of energy-wasting *syn-anti* isomerization is also inhibited, especially in five membered cyclic compounds. Padwa and Albrecht have demonstrated that the O-methyl oxime ethers of acetophenone undergo efficient photochemical *syn-anti* isomerization in both the singlet and triplet states[13].

We examined initially the photochemical reactivity of 2-ethoxypyrrolin-5-one (*12*) with olefins because of its obvious similarity to 2-cyclopentenone. Irradiation of *12* in the presence of 1,1-dimethoxyethene or cyclohexene in *tert*-butyl alcohol solvent did not give a (2 + 2)-cycloadduct but the rearrangement product *tert*-butyl N-(ethoxycyclopropyl)carbamate (*13*)[14]. The novelty and synthetic potential of the rearrangement of *12* to *13* lured us temporarily away from our original pursuit. However, as will become evident later, our studies of the rearrangement reaction were important in our eventual discovery of the (2 + 2)-photocycloaddition reaction.

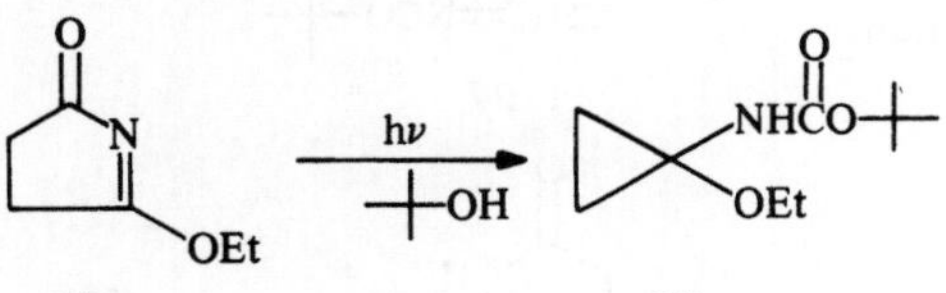

12 13

## II. Type I. Rearrangement of 2-Ethoxypyrrolin-5-ones

### 1. Synthesis of Reactants

2-Ethoxypyrrolin-5-one (*12*) was first prepared in 1891 by Comstock and Wheeler by O-alkylation of the silver salt of succinimide with ethyl iodide[15]. We have prepared *12* and various derivatives of *12* using their procedure or a modified version

**Table 1. Starting materials, ultraviolet spectral data, and photoproducts**

| Keto imino ether (isolated yield) | UV absorption nm (ε) (solvent) | Photoproduct or derivative (isolated yield) |
|---|---|---|
| 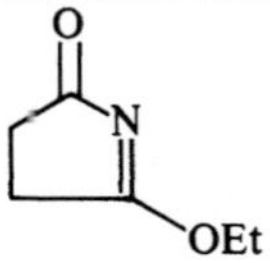 *12* (71)[1]) | 273 (55) (cyclohexane) end absorption 265 (45) (absolute EtOH) end absorption | 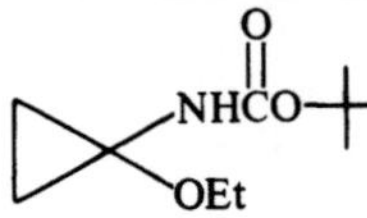 *13* (73) |
| | | 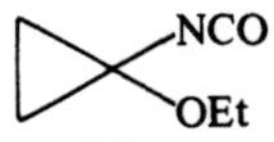 *20* (78) |
| 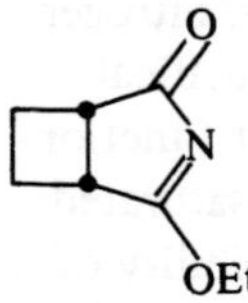 *15* (84)[1]) | 274 (64) (THF) 213 (4930) | 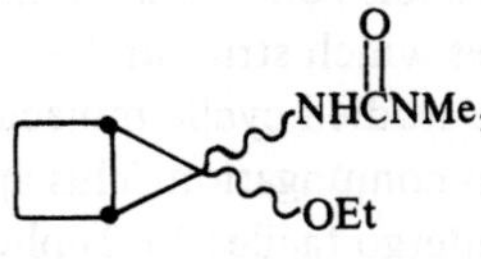 *21* (43) |
| 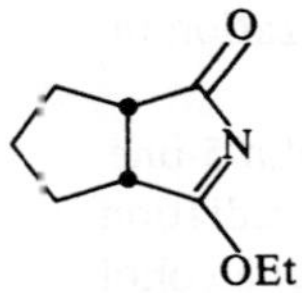 *16* (59) | 273 (50) (THF) 214 (3010) | 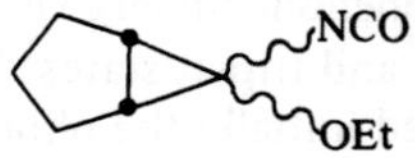 *22* (76) |
| | | 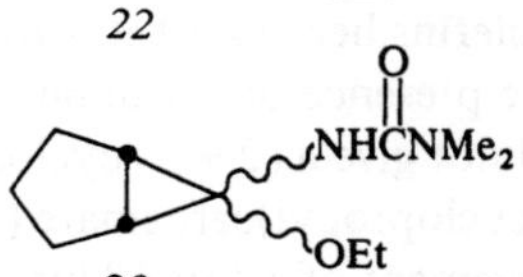 *23* (56) |
| 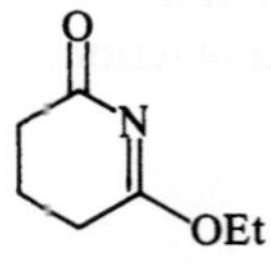 *14* (38) | 275 (60) (cyclohexane) end absorption | 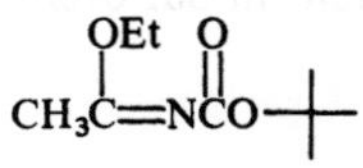 *27* (15) |
| | | 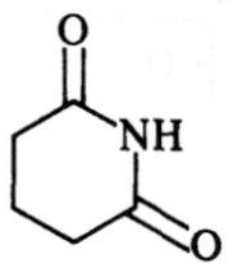 *29* (23) |
| 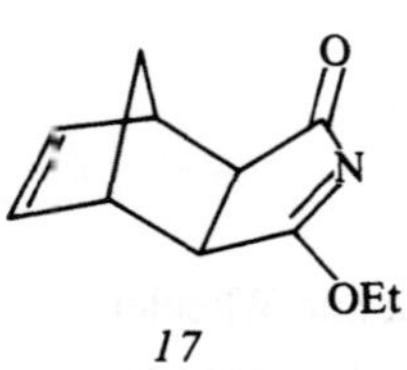 *17* (45) | 277 (60) ($CH_3CN$) | 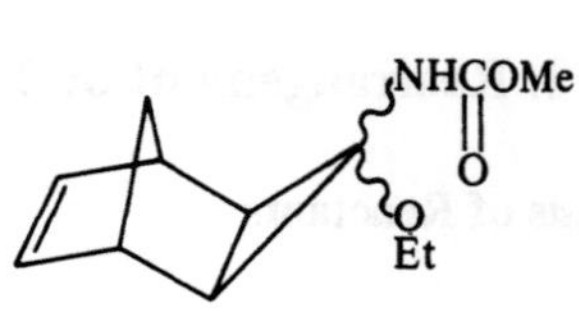 *24* (54) |

Table 1 (continued)

| Keto imino ether (isolated yield) | UV absorption nm (ε) (solvent) | Photoproduct or derivative (isolated yield) |
|---|---|---|
| 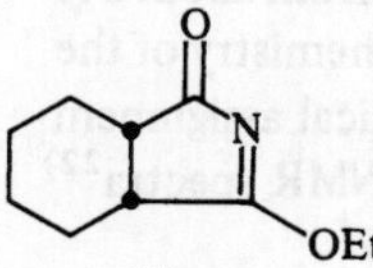 (89) *18* | 258 (50) (THF)<br>220 (2300) | 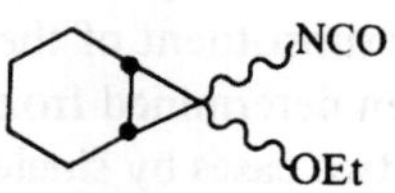 (64) *25* |
| 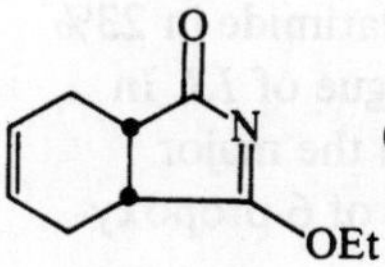 (74) *19* | 257 (64) (THF)<br>208 (6320) | 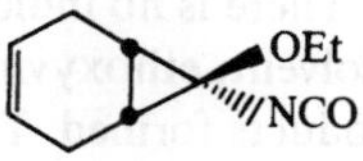 (53) *26* |

[1]) The yield was corrected for recovered starting material.

of their procedure[16, 17]. Isolated yields are in the range of 60–90% (see Table 1). 6-Ethoxy-4,5-dihydro-2(3H)-pyridone (*14*), a homologue of *12*, can similarly be prepared by O-alkylation of the silver salt of glutarimide, albeit in lower yield[18]. Triethyloxonium fluoroborate and structurally related alkylating agents work poorly or not at all in this application.

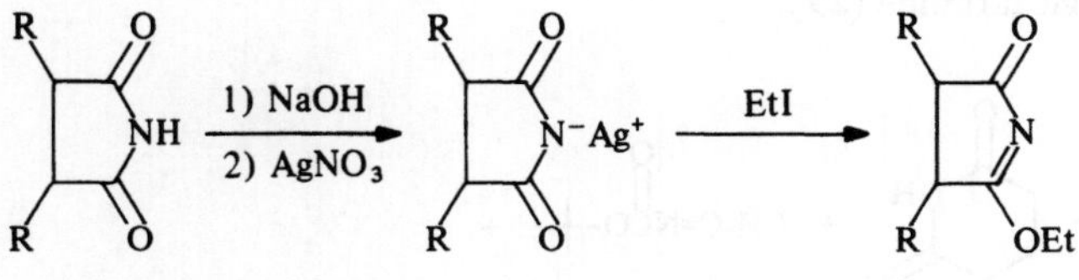

## 2. Photochemical Reactions

Irradiation of a *tert*-butyl alcohol solution of 2-ethoxy-pyrroline-5-one (*12*) with Vycor filtered light from a 450-W mercury lamp results in formation of *tert*-butyl N-(ethoxycyclopropyl) carbamate (*13*) in 73% isolated yield[17, 18]. When the irradiation is conducted in the non-hydroxylic solvent, tetrahydrofuran, 1-ethoxycyclopropyl isocyanate (*20*) can be isolated in 78% yield. Both *13* and *20* can be prepared in synthetically useful quantities.

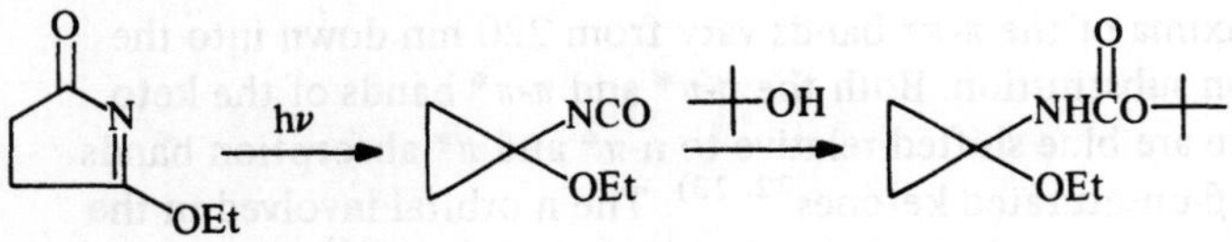

*12* *20* *13*

Four bicyclic analogues (*15, 16, 18,* and *19*) of *12* photorearrange to bicyclo (n.1.0)alkanes functionalized at the one-carbon bridge in moderate but useful yields (see Table 1)[16]. Similarly, a tricyclic system *17* photorearranges to the functionalized tricyclo[3.2.1.0$^{2,4}$] system *24*[19]. Photoproducts *21, 22, 23, 24,* and *25* are formed as a mixture of stereoisomers. The ratio of stereoisomers varies from about 3:1 to 10:1 depending on structure. For all the bicyclic systems the stereochemistry of the ethoxy substituent of the major stereoisomer is exo. This stereochemical assignment has been determined from chemical shifts in natural abundance $^{15}$N-NMR spectra[22] and in two cases by single crystal X-ray diffraction analysis[21].

Irradiation of a *tert*-butyl alcohol solution of 6-ethoxy-4,5-dihydro-2(3H)-pyridone (*14*) under conditions similar to those described for *12* gives two products, *tert*-butyl N-(ethoxyethylidene)carbamate (*27*) in 15% yield and glutarimide in 23% yield[18]. There is no indication of formation of the cyclobutyl analogue of *13*. In aprotic solvent, ethoxyvinyl isocyanate (*28*) and glutarimide (*29*) are the major photoproducts formed. The gaseous byproducts from the irradiation of 6-propoxy-

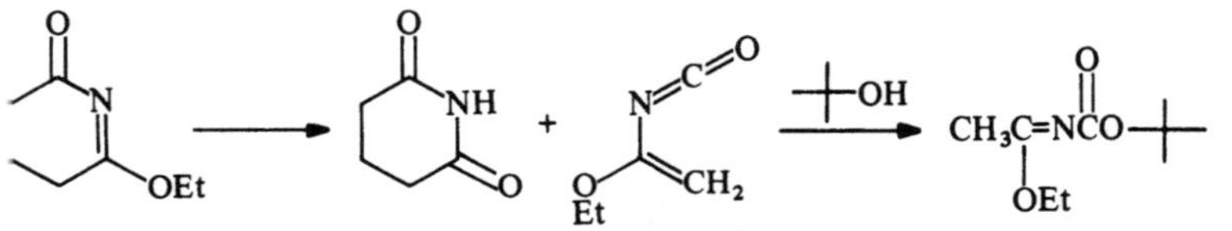


*14* *29* *28* *27*

4,5-dihydro-2(3H)-pyridone in *tert*-butyl alcohol solvent have been trapped and identified as ethene and propene by mass spectrometry. The non-gaseous products from the irradiation of the propoxydihydropyridone are *tert*-butyl N-(propoxy-ethylidene)carbamate and glutarimide (*29*).

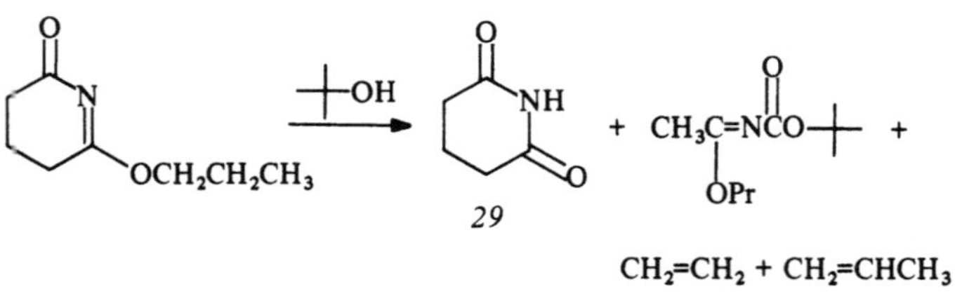


### 3. Spectroscopic and Quantum Yield Measurements

Ultraviolet absorption data for all the molecules bearing the keto imino ether chromophore are shown in Table 1. The longest wavelength absorption has been assigned to an n-$\pi^*$ band consistent with the low extinction coefficients and the solvent shift observed for *12*. The maxima of the $\pi$-$\pi^*$ bands vary from 220 nm down into the vacuum UV depending on substitution. Both the n-$\pi^*$ and $\pi$-$\pi^*$ bands of the keto imino ether chromophore are blue shifted relative to n-$\pi^*$ and $\pi^*$ absorption bands of structurally related $\alpha,\beta$-unsaturated ketones[22, 23]. The n orbital involved in the n-$\pi^*$ transition has not been assigned. Extended Hückel calculations[24] of *12* indicate that the highest occupied orbital in the ground state is a linear combination of atomic orbitals in the σ-plane with significant coefficients for formally n orbitals on the carbonyl oxygen and the nitrogen.

The quantum yields of formation of *tert*-butyl N-(ethoxycyclopropyl)carbamate (*13*) from irradiation of *12* and of *tert*-butyl N-(ethoxyethylidene)carbamate (*27*) and glutarimide (*29*) from irradiation of *14* are moderately high and are shown in Table 2. The effect of the triplet quencher *cis*-piperylene on the quantum yields has been examined. Over the concentration range 0.01–0.10M *cis*-piperylene, no appreciable effect is observed as indicated by an average Stern-Volmer slope of 1.3 l/mole. Energy transfer from triplet *12* and *14* does occur since *trans*-piperylene is formed during the irradiations. Triplet counting plots of reciprocal of *cis-trans* isomerization quantum yield versus reciprocal of *cis*-piperylene concentration are linear and yield the intersystem crossing efficiencies and triplet lifetimes shown in Table 2. Acetone sensitization of the photorearrangement of *12* has also been examined. Over the concentration range 0.1–1.0 M acetone, no sensitization occurred. The spectroscopic, quenching, and sensitization data have been interpreted in terms of reaction *via* an n, π* singlet state and intersystem crossing with moderate efficiency to an unreactive triplet state, probably also of n, π* configuration (*vide infra*). A kinetic mechanism is shown in Scheme 1.

Table 2. Quantum yield data for photoreactions of *12* and *14*

| Compd. | Quantum Yield (Carbamate) | Quantum Yield (Imide) | Stern-Volmer slope ($M^{-1}$) | $\phi$ isc | $k_q\tau_t$ | $\tau_t$ |
|---|---|---|---|---|---|---|
| 12 | 0.31 | | 1.4 | 0.19 | 60 | $3 \times 10^{-8}$ |
| 14 | 0.067 | 0.025 | 1.2 | 0.10 | 49 | $2.5 \times 10^{-8}$ |

$K + h\nu \longrightarrow {}^1K^*$
${}^1K^* \longrightarrow K$
${}^1K^* \longrightarrow$ products
${}^1K^* \longrightarrow {}^3K^*$
${}^3K^* \longrightarrow K$
${}^3K^* + Q \longrightarrow K + {}^3Q^*$
K ≡ keto imino ether
Q ≡ quencher

Scheme 1

## 4. Stereochemistry

The stereochemistry of the rearrangement of 2-ethoxypyrrolin-5-one to N-(ethoxycyclopropyl)isocyanate at carbons 2 and 4 has been investigated using *cis*- and *trans*-3,4-dimethyl-2-ethoxypyrrolin-5-ones (*30* and *31*, respectively). Irradiation of *30* in tetrahydrofuran solvent followed by reaction with anhydrous methanol gives almost exclusively two diastereoisomeric methyl N-(*cis*-2,3-dimethyl-l-ethoxycyclopropyl)carbamates (*32a* and *32b*) in a ratio of 3.3:1.0. Irradiation of *31* under similar conditions followed by reaction with methanol gives almost exclusively methyl N-(*trans*-2,3-dimethyl-l-ethoxycyclopropyl)carbamate *33*. Both pyrrolinones *30* and *31* rearrange with 98 ± 1% retention of stereochemistry and 2 ± 1% inversion of stereochemistry at carbon 4. The rearrangement of *cis*-dimethylpyrrolinone *30* occurs with stereoselectivity at carbon 2 similar to that observed with the bicyclic systems *15, 16, 18,* and *19.* The stereochemical assignments were made from relative chemical shifts in natural abundance $^{13}C$ and $^{15}N$-NMR spectra.

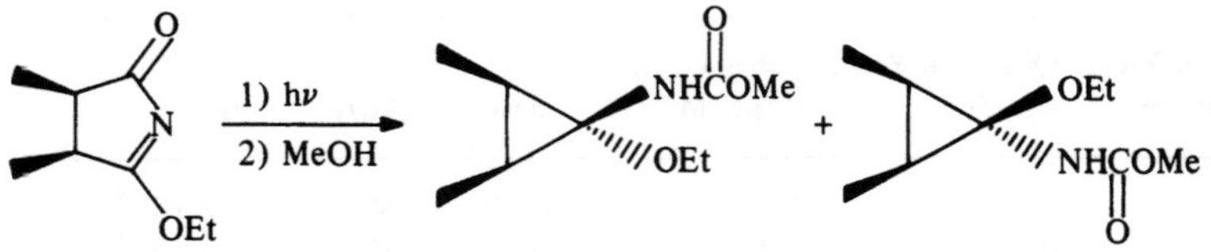


*30* *32a* *32b*

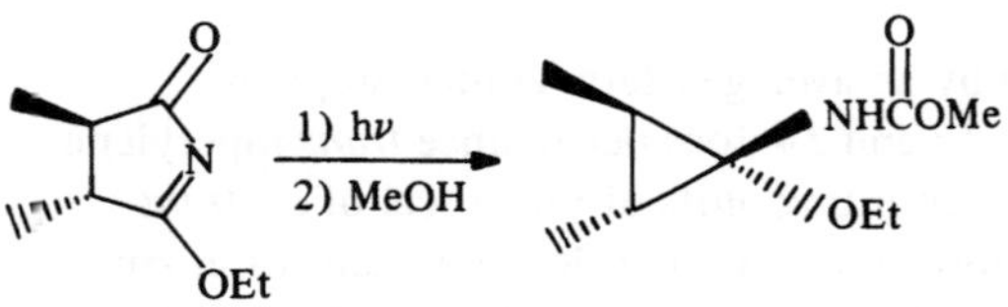


*31* *33*

## 5. Mechanistic Discussion

There are two reasonable interpretations of the stereospecificity at carbon 4 of the pyrrolinones (*30* and *31*). The reaction could be completely stereospecific and occur by a symmetry-allowed concerted [$\sigma 2s + \pi 2s$] pathway (Scheme 2)[26], equivalently *via* a Hückel four-electron transition state[22], or *via* a short-lived singlet diradical intermediate such as *34* which ring closes by a rate more than an order of magnitude faster than the rate for bond rotation (Scheme 3). The stereoselectivity at carbon 2 of the pyrrolinones *30* and *31* and at the equivalent carbons of the bicyclic and tricyclic systems (*15–19*) is probably governed by steric interactions in the transition states for reaction *via* either mechanism.

The concerted pathway is analogous to the mechanism proposed for the singlet-state 1,3-sigmatropic rearrangement of $\beta,\gamma$-unsaturated ketones[28], and the diradical mechanism is analogous to the diradical mechanism proposed for the photochemical ring expansion of cyclobutanones[29–31]. Both of these reactions also show a high degree of stereospecificity[28, 29, 31]. Because the photorearrangement of *30* and *31* does not appear to be completely stereospecific and because the orbitals involved are approximately orthogonal, at least in the ground state, we tend to favor a mechanism with a diradical intermediate (Scheme 3).

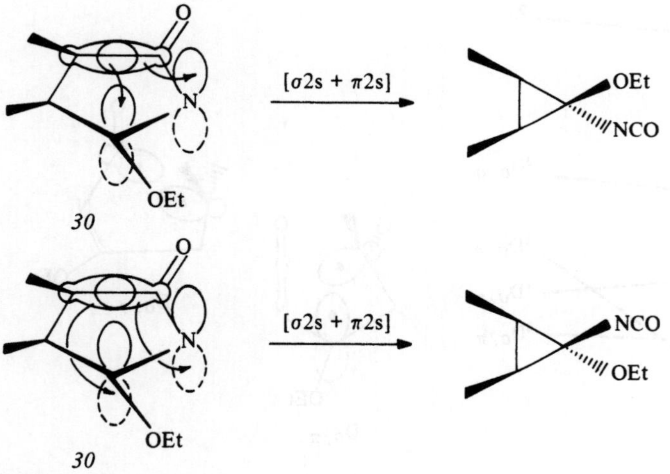


Scheme 2

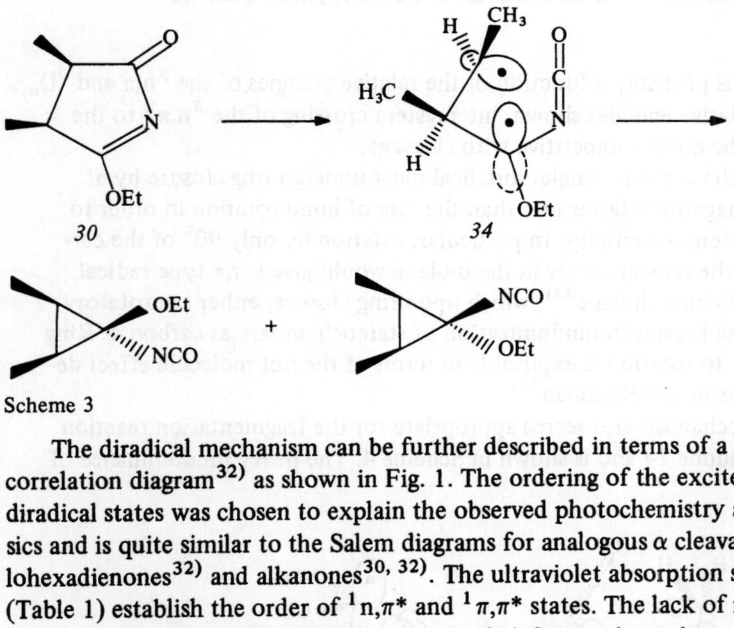


Scheme 3

The diradical mechanism can be further described in terms of a Salem state correlation diagram[32] as shown in Fig. 1. The ordering of the excited states and diradical states was chosen to explain the observed photochemistry and photophysics and is quite similar to the Salem diagrams for analogous α cleavage of 2,4-cyclohexadienones[32] and alkanones[30, 32]. The ultraviolet absorption spectral data (Table 1) establish the order of $^1n,\pi^*$ and $^1\pi,\pi^*$ states. The lack of reactivity of *1* in the (2 + 2)-photocycloaddition reaction of olefins to the carbon-nitrogen double bond of *1* is consistent with an n,π* configuration for the lowest energy triplet state (*vide infra*). The σ,π diradical states were placed below the σ,σ diradical states because of π-delocalization effects in the former[32]. With this arrangement of states (Fig. 1), the n,π* singlet state smoothly correlates with the singlet σ,π diradical state. Inefficiency in the singlet manifold can possibly be explained by the crossing of the $^1n,\pi^* \rightarrow {}^1D_{\sigma,\pi}$ potential energy surface with the surface connecting the ground state with $D_{\sigma,\sigma}$. Decay of the n,π* singlet state by a mechanism of this type should give starting material without a change in stereochemistry as observed[20, 33]. The reactivity

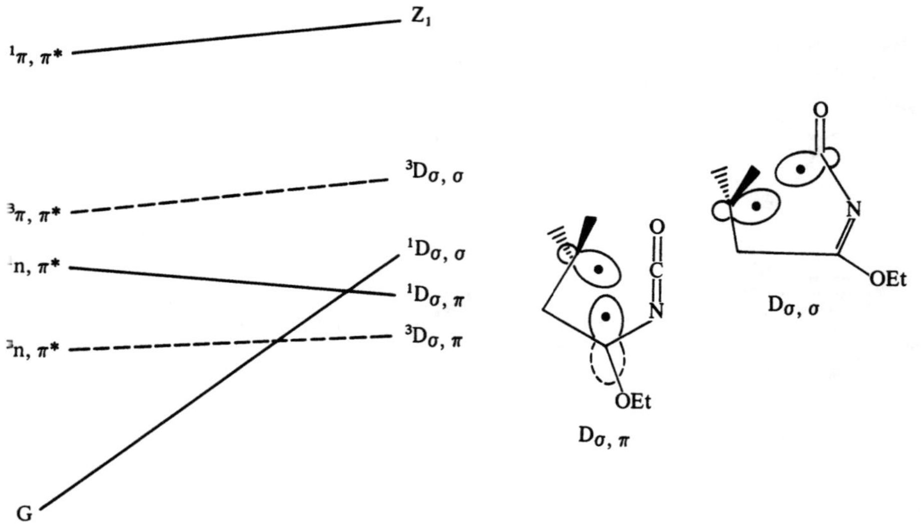


Fig. 1. State correlation diagram for the α-cleavage of 2-ethoxypyrrolin-5-one (*12*)

of the $^3n,\pi^*$ state is probably a function of the relative energies of the $^3n,\pi$ and $^3D_{\sigma,\pi}$ states. With the relative energies shown, intersystem crossing of the $^3n,\pi^*$ to the ground state may be quite competitive with cleavage.

Once formed the σ,π type singlet diradical must undergo ring closure by at least an order of magnitude faster rate than the rate of bond rotation in order to give the observed stereospecificity. In particular, rotation by only 90° of the carbon of *34* bearing the radical center in the σ-plane would give a π,π type radical, also known as a 0,0-trimethylene[33], which upon ring closure, either conrotatory or disrotatory, would result in randomization of stereochemistry at carbon 4. Ring closure of *34* prior to rotation is explicable in terms of the hot molecule effect described by Stephenson and Brauman[34].

A diradical mechanism also seems appropriate for the fragmentation reaction of the dihydropyridone *14* and is shown in Scheme 4. The heavy predominance of

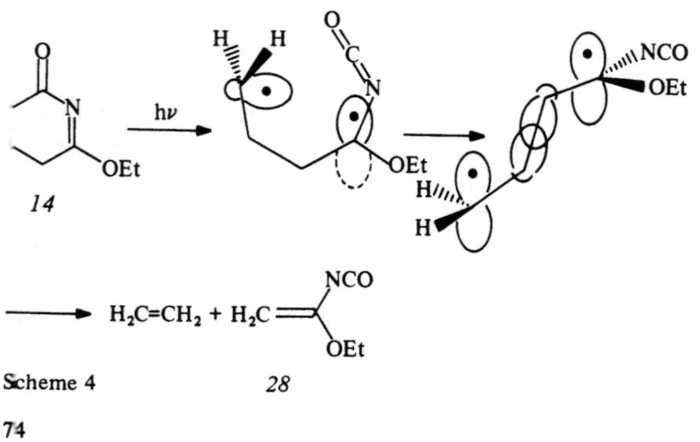


Scheme 4

fragmentation over cyclization suggests that rotation about sigma bonds occurs in order to achieve a conformation which favors bond cleavage of the diradical intermediate[35]. Formation of glutarimide from irradiation of *14* and 6-propoxy-4,5-dihydro-2(3H)-pyridone is rationalized in terms of a Norrish Type II reaction involving a carbon-nitrogen double bond. The only definitive evidence is formation of propene from irradiation of the propoxydihydropyridone.

## 6. Carbocyclic Analogues and Related Systems

α-Cleavage is an uncommon reaction of α, β-unsaturated ketones structurally related to *12* and *14*. Photodimerization and photocycloaddition to olefins are generally the observed reactions. There are, however, several examples in the literature of related systems which α-cleave. 2-Cyclopentenones substituted in the 5 position (*35*) preferentially α-cleave in dilute solution to give cyclopropy ketenes (*36*) among other products as observed by Agosta, Kende, and co-workers[36]. These reactions have been shown to occur under sensitized conditions, but the details of the mechanism are still unknown.

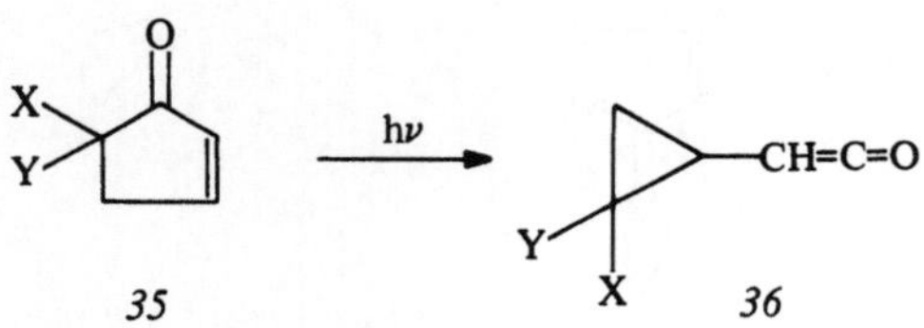


*35* *36*

X = Y = Ph
X = Y = $CH_3$
X = $C_2H_5$; Y = H
X = n–$C_3H_7$; Y = H
X = $OC_2H_5$; Y = H

McMurry and co-workers have reported that irradiation of the epimeric hydroxycyclopentenones *37a, b* yields the stereoisomeric lactones *38a, b* most likely via an intermediate hydroxycyclopropyl ketene[37]. The stereospecificity of the reaction may result from hydrogen bonding between the hydroxyl group and the incipient carbonyl in the transition state. The results of quenching and sensitization experiments implicate a reactive singlet state.

An intermediate epoxy ketene (*39*) from α-cleavage of 2,5-diphenyl-3(2H)-furanone (*40*) has been proposed by Padwa and co-workers to explain photoisomerization to 4,5-diphenyl-2(5H)-furanone (*41*)[38]. The epoxy ketene was not observed when the irradiation was monitored by infrared spectroscopy and was not trapped by methanol. The authors suggest that the intermediate may be formed with excess vibrational energy and as a result undergo very rapid intramolecular reaction.

Herz and Iyer have proposed an α-cleavage mechanism with the intermediacy of a cyclopropyl ketene for formation of the polycyclic acetal *42* from photolysis of the rigid cyclopentenone *43*[39]. The cleavage reaction is thought to occur *via* an n,π* triplet state of unusually high energy for a cyclopentenone (75 kcal/mole).

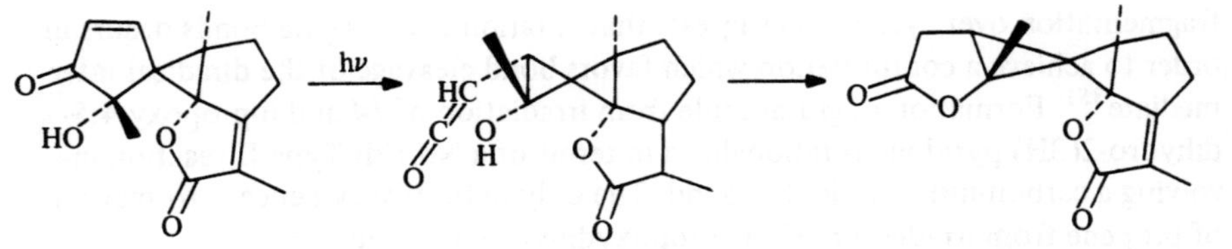


*37a* *38a*

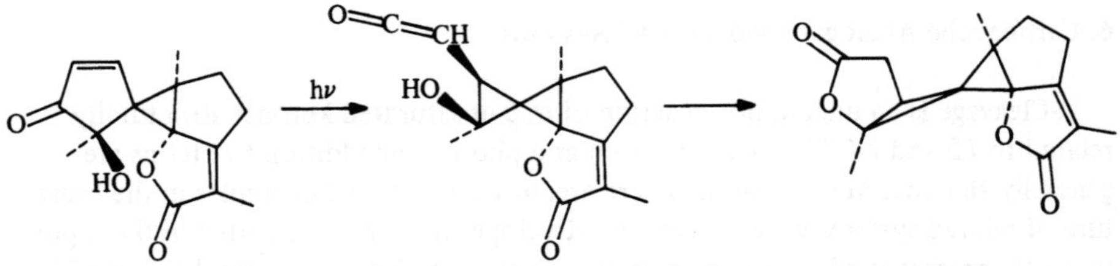


*37b* *38b*

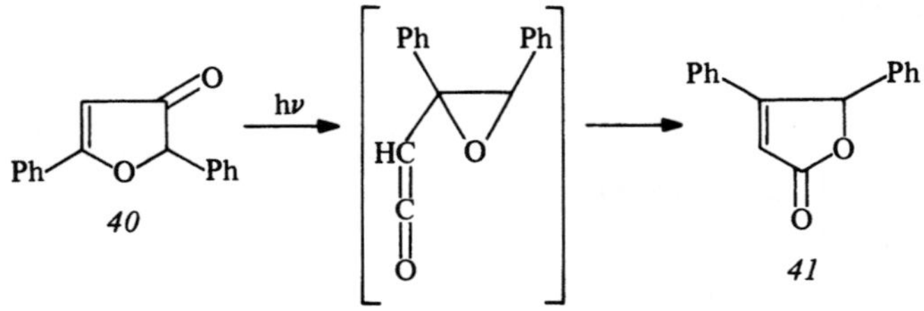


*40* *41* *39*

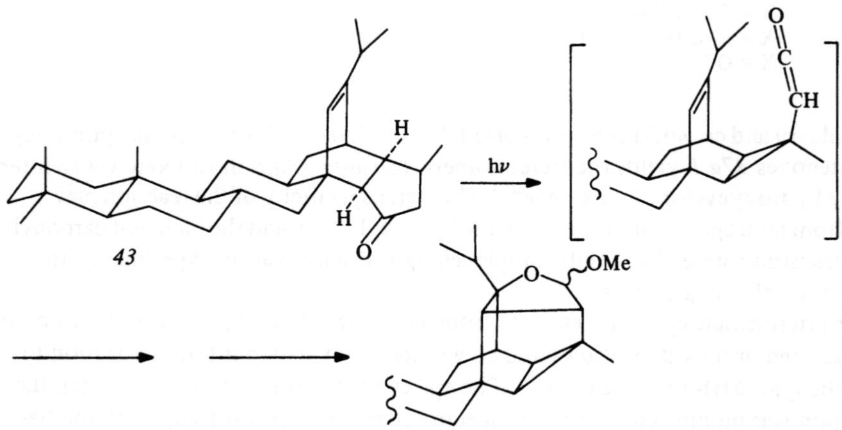


*43* *42*

A number of bicyclic ketones such as bicyclo[3.2.1]octa-3,6-dien-2-one which are both $\alpha,\beta$ and $\gamma',\delta'$ unsaturated undergo photoisomerization to vinylcyclopropyl ketenes[40]. These reactions may be mechanistically related to the reactions of *12* and *14*; however, the reactions have also been described as [3,3]-sigmatropic shifts[41].

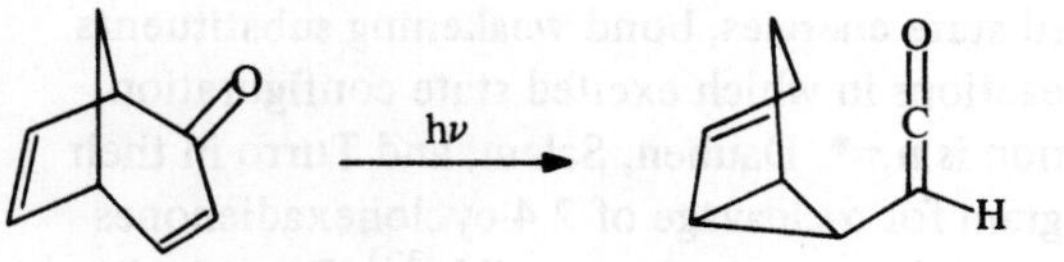


Of the many 2-cyclohexenones that have been irradiated only some of the 6-hydroxy-2-cyclohexenones α-cleave. A hydroxycyclobutyl ketene is proposed as an intermediate in the photochemical rearrangement of 4,4-dimethyl-6-hydroxy-2-cyclohexenone (*44*) to the bicyclic lactone *45*. The hydroxy substituent is thought to facilitate the α-cleavage by stabilizing the intermediate diradical and by imparting rigidity to the ring system through intramolecular hydrogen bonding with the carbonyl.

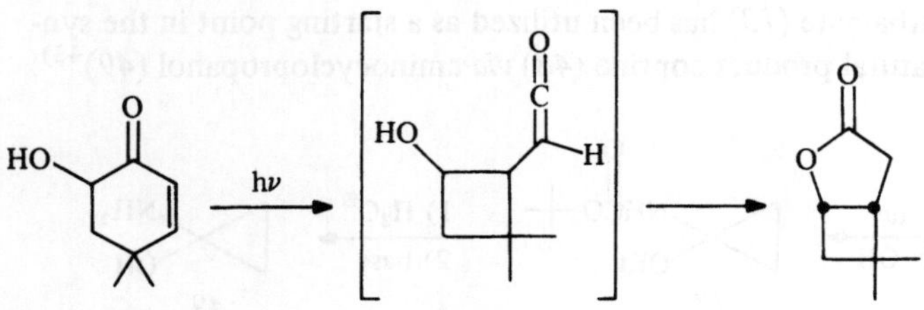


*44* *45*

α-Cleavage is the predominant mode of photochemical reactivity of 2,4-cyclohexadienones[42]. A σ,π type diradical has been proposed as an intermediate[32].

α-Cleavage has also been reported for a seven-membered ring compound electronically related to *12* and *14*. Irradiation of 3,6-dihydro-3,3,6,6-tetramethyl-2H-azepin-2-one (*46*) in hexane gives in 20% yield *cis*-2,2-dimethyl-3-isobutenylcyclopropyl isocyanate (*47*)[43].

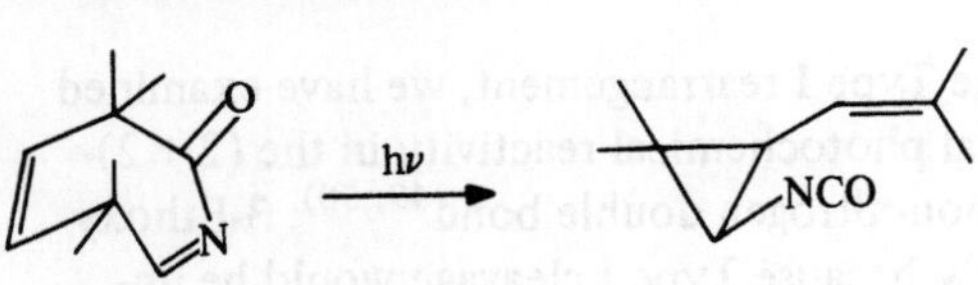


*46* *47*

In comparing these various α-cleavage reactions of unsaturated ketones, the structural and electronic factors which are apparently important for this mode of photochemical reactivity are excited state energy, excited state configuration, strength of the sigma bond α to the carbonyl, and rigidity of the ring. If the excited state energy is high as with the keto imino ethers *12* and *14*, α-cleavage occurs in the absence of bond weakening substituents. With the cyclopentenones and cyclohexenones, which have lower excited state energies, bond weakening substituents appear to be necessary. For those reactions in which excited state configuration has been established, the configuration is n,π*. Dauben, Salem, and Turro in their analysis of the state correlation diagram for α-cleavage of 2,4-cyclohexadienones indicate that reactivity from the π,π* triplet state is also possible[32]. Ring rigidity is probably an especially important factor for the α-cleavage of 2-cyclohexenones which are inherently less rigid than 2-cyclopentenones and can potentially dissipate electronic energy by *cis-trans* isomerization[44].

### 7. Synthetic Applications

The photochemical rearrangement of 2-ethoxypyrrolin-5-one (*12*) to *t*-butyl N-(ethoxycyclopropyl)carbamate (*13*) has been utilized as a starting point in the synthesis of the unusual natural product coprine (*48*) *via* aminocyclopropanol (*49*)[45].

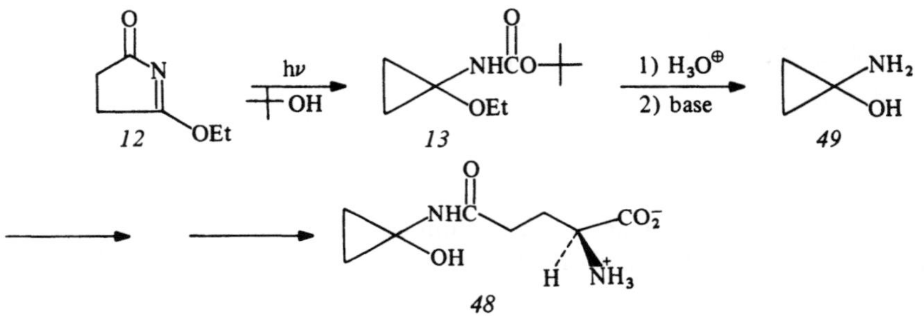


Although no additional synthetic applications have been completed at this time, cyclopropanes *13*, *23*, and *49* are derivatives of cyclopropanone, and derivatives of this type have been used as precursors for a variety of compounds[46] such as β-lactams[47], cyclobutanones[48], and cyclopropanols[47].

## III. (2 + 2)-Photocycloaddition to the Carbon-Nitrogen Double Bond

Concurrent with our development of the Type I rearrangement, we have examined 3-ethoxyisoindolenone (*50*) for potential photochemical reactivity in the (2 + 2)-photocycloaddition reaction to the carbon-nitrogen double bond[49, 50]. 3-Ethoxyisoindolenone (*50*) was selected for study because Type I cleavage would be unfavorable, resulting in formation of a high-energy phenyl radical, and (2 + 2)-photocycloaddition might be competitive.

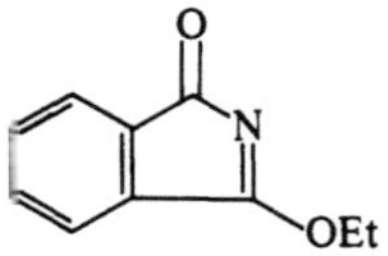


*50*

## 1. Photochemical Reactions of 3-Ethoxyisoindolenone

Irradiation of *50* with a 450-W mercury lamp through a Corex filter in acetonitrile or *t*-butyl alcohol solvent does not result in formation of products from α-cleavage. 3-Ethoxyisoindolenone is slowly destroyed by the irradiation, and high molecular weight unidentified products are formed.

When *50* is similarly irradiated through a Pyrex filter in *t*-butyl alcohol, benzene, or methylene chloride solvent in the presence of 4 equiv. of 1,1-dimethoxyethene, a (2 + 2)-photocycloadduct (*51*) is formed regiospecifically in 50% isolated yield. No reaction is observed when *50* and 1,1-dimethoxyethene are allowed to stand in the dark for an equivalent period of time. Irradiation of *50* with 4 equiv. of cyclohexene in methylene chloride solvent gives a 51% isolated yield of the (2 + 2)-cycloadduct *52* assigned *cis-anti-cis* stereochemistry. These two cycloadditions were the first two well defined examples of the (2 + 2)-photocycloaddition reaction of an olefin to a carbon-nitrogen double bond. With furan the (2 + 2)-photocycloadduct *53* is formed, again regiospecifically.

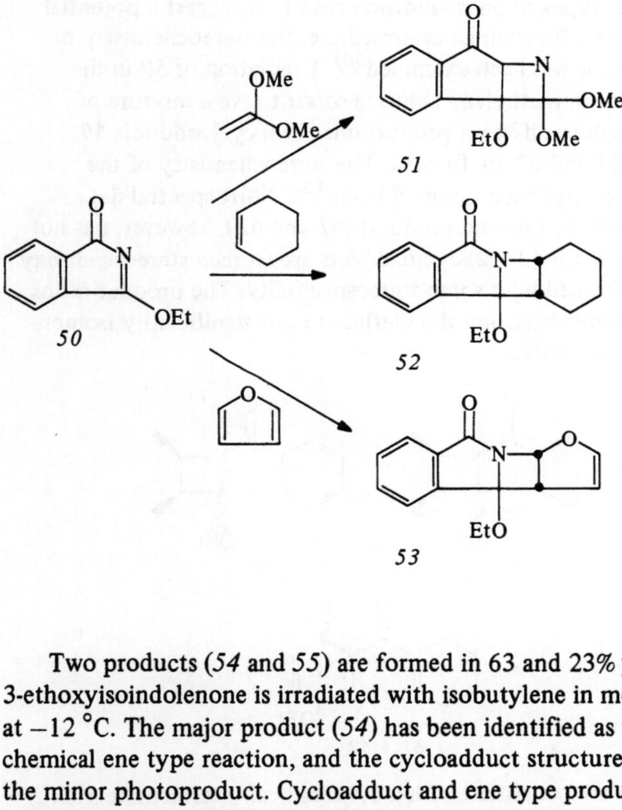


*51*

*50*

*52*

*53*

Two products (*54* and *55*) are formed in 63 and 23% yields, respectively, when 3-ethoxyisoindolenone is irradiated with isobutylene in methylene chloride solvent at −12 °C. The major product (*54*) has been identified as the product of a photochemical ene type reaction, and the cycloadduct structure *55* has been assigned to the minor photoproduct. Cycloadduct and ene type products (*56* and *57*) are also formed in 25 and 23% yields, respectively, when *50* is irradiated in the presence of tetramethylethylene. Additionally an unprecedented benzoazepinone (*58*) is formed in 12% yield.

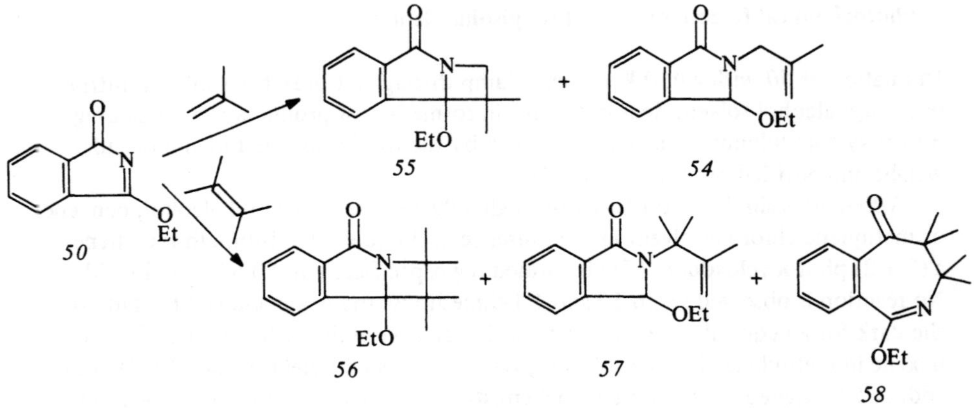

*50*, *55*, *54*, *56*, *57*, *58*

## 2. Stereochemistry of Photoaddition to 3-Ethoxyisoindolenone

The structures of the products formed in the photoadditions to 3-ethoxyisoindolenone and precedent from other types of photoaddition reactions suggest a potential 1,4-diradical intermediate. To test for such an intermediate, the stereochemistry of addition to *cis*- and *trans*-2-butene has been examined[50]. Irradiation of *50* in the presence of *cis* or *trans*-2-butene in methylene chloride solvent gave a mixture of the same four products, although in different proportions. Two cycloadducts *59* and *60* and two ene products *61* and *62* are formed. The stereochemistry of the methyl substituents of *59* and *60* has been assigned from $^{13}$C NMR spectral data. A stereochemical assignment for the two ene products (*61* and *62*), however, has not been made. As shown in Table 3, the (2 + 2)-cycloadducts are formed stereorandomly and the two ene products are formed with some stereospecificity. The product ratios are independent of olefin concentration, and the olefins are not significantly isomerized during the course of the irradiations.

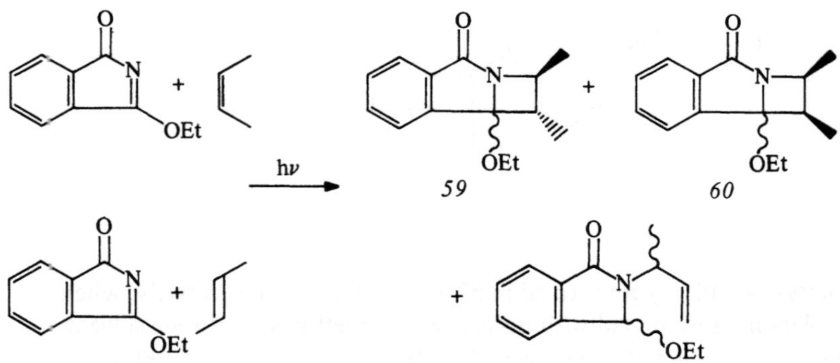

*59*, *60*, *61 + 62*

The stereochemical results are consistent with a 1,4-diradical intermediate such as *63* with a lifetime sufficiently long to permit loss of stereochemistry by rotation for the cycloaddition reaction. At most, however, the same diradical can be an intermediate in only a portion of the ene reaction.

Table 3. Product ratios from photoaddition of *50* to *cis* and *trans*-2-butene

| Olefin | *59/60* | *61/62* | *(59 + 60) / (61 + 62)* |
|---|---|---|---|
| *cis*-2-butene | 2.0 | 3.1 | 0.72 |
| *trans*-2-butene | 2.1 | 0.69 | 1.8 |

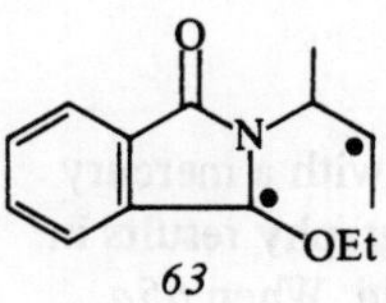

*63*

We will return to a possible explanation of the stereochemical results after a slight digression into a complementary system.

## 3. Photochemical Reactions of 2-Aryl-2-oxazolin-4-ones

Subsequent to our discovery of the photocycloaddition reactions of 3-ethoxyisoindolenone (*50*), we studied the photochemical reactivity of a series of 2-aryl-2-oxazolin-4-ones (*64*)[51, 52]. This system was selected for investigation in order to link our understanding of the photochemical α-cleavage reaction of 2-ethoxypyrrolin-5-one (*12*) with our understanding of the photoaddition reactions of *50*. As described earlier the benzo substituent of *50* precludes the α-cleavage reaction presumably by strengthening the carbon-carbon single bond α to the carbonyl. The benzo substituent, however, also changes the chromophore of *50* relative to the chromophore of *12* since it is conjugated with the keto imino ether. This is evident from the dramatic differences in the UV absorption spectra of *12* and *50* (*vide infra*). The photochemistry of *64* has been examined with the idea of determining the effect of the additional conjugation of the keto imino ether chromophore with the benzo group of *50* on reactivity in photoaddition. The oxazolinones *64* are keto imino ethers which have extended conjugation with an aryl group but which are also structurally capable of α-cleavage. The photochemical reactivity of 2-phenyl-2-oxazolin-4-one *65a* has been explored, and the effect of substituents on the phenyl group has been studied in order to make a configurational assignment for the reactive states.

The requisite 2-aryl-2-oxazolin-4-ones can be synthesized using the procedure briefly described by Gordon for the preparation of 2-phenyl-2-oxazolin-4-one (*65a*)[53]. The method utilizes an intramolecular O-alkylation of an imide ambident anion.

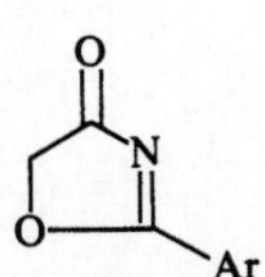

*64*

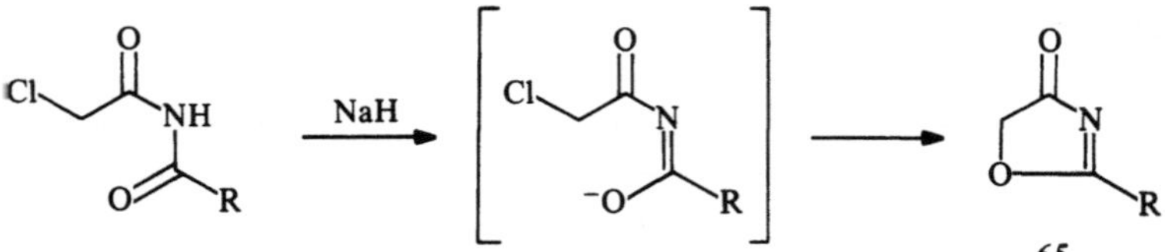


*65*

a, R = phenyl
b, R = *p*–methoxyphenyl
c, R = *m*–methoxyphenyl
d, R = *m*–trifluoromethylphenyl

Irradiation of 2-phenyl-2-oxazolin-4-one (*65a*) in glyme solvent with a mercury lamp through a Corex filter followed by silica gel column chromatography results in the formation of 4-phenyl-4-oxazolin-2-one (*66*) in 45% isolated yield. When *65a* is similarly irradiated in glyme or benzene solvent followed by solvent evaporation and molecular distillation, an unstable product is isolated and identified as 2-phenyl-2-oxiryl isocyanate (*67a*). Treatment of the isocyanate with silica gel results in rearrangement to 4-phenyl-4-oxazolin-2-one (*66*). Similarly 2-(*m*-trifluoromethyl-phenyl)-2-oxazolin-4-one (*65d*) rearranges upon irradiation to the corresponding isocyanate, 2-(*m*-trifluoromethyl-phenyl)-2-oxiryl isocyanate (*67d*). Isocyanate *67d* does not rearrange upon silica gel chromatography. The methoxy-substituted oxazolinones *65b* and *65c* do not rearrange to oxiryl isocyanates when irradiated under identical conditions.

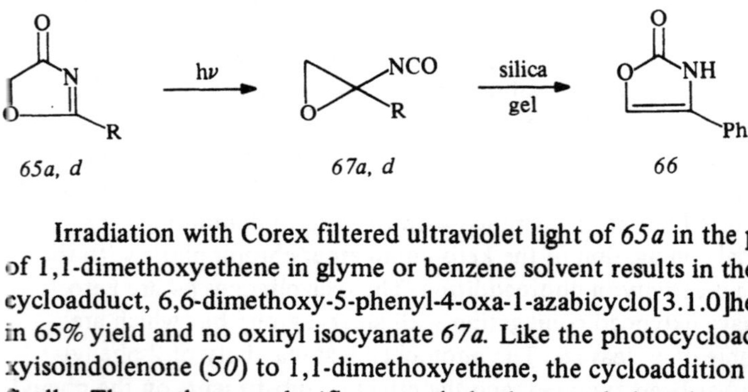


*65a, d* *67a, d* *66*

Irradiation with Corex filtered ultraviolet light of *65a* in the presence of 20 equiv. of 1,1-dimethoxyethene in glyme or benzene solvent results in the formation of a cycloadduct, 6,6-dimethoxy-5-phenyl-4-oxa-1-azabicyclo[3.1.0]heptan-2-one (*68a*), in 65% yield and no oxiryl isocyanate *67a*. Like the photocycloaddition of 3-ethoxyisoindolenone (*50*) to 1,1-dimethoxyethene, the cycloaddition occurs regiospecifically. The methoxy and trifluoromethyl substituted phenyloxazolinones *65b, c,* and *d* similarly photocycloadd regiospecifically to 1,1-dimethoxyethene.

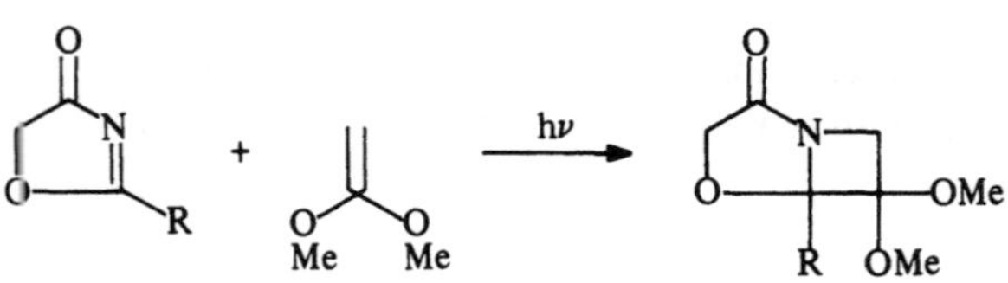


*65a, b, c, d* *68a, b, c, d*

a, R = phenyl
b, R = *m*–methoxyphenyl
c, R = *p*–methoxyphenyl
d, R = *m*–trifluoromethylphenyl

Regiospecific cycloaddition of *65a, b,* and *c* to furan has also been observed. None of the oxazolinones give photoadducts with other olefins, including methyl vinyl ether, cyclohexene, *cis*-2-butene, *trans*-2-butene, and styrene.

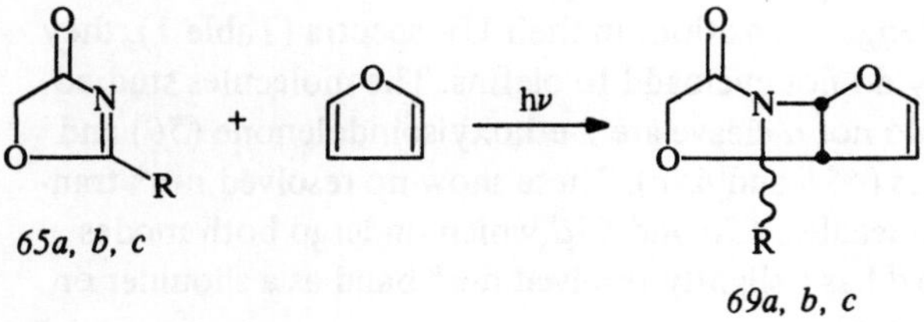

*65a, b, c*

*69a, b, c*

## 4. Spectroscopic Measurements

3-Ethoxyisoindolenone (*50*) and the 2-aryl-2-oxazolin-4-ones *65a, 65b, 65c,* and *65d* have UV absorption maxima which are contrasted with the UV absorption maximum of 2-ethoxypyrrolin-5-one (*12*) in Table 4[52]. The UV absorption spectra of *50, 65a, 65b, 65c,* and *65d* in general show intense $\pi$-$\pi^*$ bands. Only *m*-trifluoromethylphenyloxazolinone *65d* has what appears to be a slightly resolved n-$\pi^*$ band as a shoulder on the $\pi$-$\pi^*$ transition.

Table 4. UV data

| Compound | Solvent | UV $\lambda_{max,nm}$ | $\epsilon$ |
|---|---|---|---|
| *12* | Cyclohexane | 273 | 55 |
| *50* | Cyclohexane | 216 | 38,000 |
| | | 288 | 1,900 |
| | | 298 | 2,100 |
| *65a* | 95% Ethanol | 263 | 20,800 |
| *65b* | 95% Ethanol | 223 | 14,000 |
| | | 300 | 32,000 |
| *65c* | 95% Ethanol | 220 | 18,700 |
| | | 268 | 18,800 |
| | | 318 | 3,700 |
| *65d* | Glyme | 256 | 16,000 |
| | | 320 (shoulder) | 100 |

In an isopropyl alcohol-diethyl ether glass at 77 K, 3-ethoxyisoindolenone (*50*) gives a phosphorescence emission with maximum at 502 nm and lifetime of 29.7 ± 0.2 msec.[54]. In a methylcyclohexane diethyl ether glass emission at the same wavelength is observed with a lifetime of 13.6 ± 0.1 msec. The emission lifetimes are within the range characteristic of a $\pi$-$\pi^*$ configuration. A tentative assignment of the O-O band gives a triplet energy of 62 kcal/mole. No phosphorescence has been observed for the oxazolinones.

## 5. Discussion of Substituent Effects

There is a correlation between reactivity of the pyrrolinones, 3-ethoxyisoindolenone and the oxazolinones and their UV spectra[52]. The pyrrolinones all have distinct n-$\pi^*$ bands as the longest wavelength transitions in their UV spectra (Table 1), they all react by $\alpha$-cleavage, and they do not cycloadd to olefins. The molecules studied which cycloadd to olefins and do not $\alpha$-cleave are 3-ethoxyisoindolenone (*50*) and the methoxyphenyloxazolinones (*65b* and *65c*). These show no resolved n-$\pi^*$ transitions (Table 4). Of the two molecules, *65a* and *65d* which undergo both modes of photochemical reactivity, *65d* has a slightly resolved n-$\pi^*$ band as a shoulder on the intense $\pi$-$\pi^*$ transition (Table 4).

In comparing the UV spectra of *65a, b, c,* and *d*, we are examining the effect of electron-donating and electron-with-drawing substituents. The electron-donating methoxy substituent lowers the energy of the $\pi$-$\pi^*$ transition relative to the n-$\pi^*$ transition, and the electron-withdrawing trifluoromethyl substituent appears to lower the energy of the n-$\pi^*$ transition, relative to the $\pi$-$\pi^*$ transition. A similar effect of substituents on the relative energies of n,$\pi^*$ and $\pi,\pi^*$ states of aromatic ketones has been observed and correlated with reactivity in photoreduction and Norrish Type II cleavage reactions[55–58].

A conclusion which can be drawn from the correlation of keto imino ether reactivity with UV absorption is that n,$\pi^*$ character is required of the reactive excited state for $\alpha$-cleavage and $\pi,\pi^*$ character is required of the reactive excited state for (2 + 2)-photocycloaddition. Since the multiplicity of the excited states responsible for photocycloaddition to olefins is triplet (*vide infra*), the validity of a correlation with UV absorption may be questionable. However, singlet-triplet energy splittings for $\pi,\pi^*$ states are generally larger than for n,$\pi^*$ states. Therefore, the lowest triplet states of *50, 65a, 65b* and *65c* are probably of $\pi,\pi^*$ configuration. The assignment of the $\pi,\pi^*$ configuration for the lowest triplet state of *50* is also consistent with the phosphorescence lifetime (*vide supra*).

## 6. Quantum Yield Measurements of Photoaddition to 3-Ethoxyisoindolenone

The photoreactions of 3-ethoxyisoindolenone with 1,1-dimethoxyethene, tetramethylethylene, and *cis*-2-butene in methylene chloride solvent were chosen for quantum yield studies[50, 59]. With 0.06 M 1,1-dimethoxyethene the quantum yield of cycloadduct formation is 0.72, and with 0.06 M tetramethylethylene the total quantum yield of adduct formation is 0.59. Over the concentration range 0.06–2.00 M, the quantum yields of product formation from addition to 1,1-dimethoxyethene and tetramethylethylene are inversely related to the olefin concentration. Plots of reciprocal of quantum yield of product formation vs. reciprocal of olefin concentration one non-linear; however, plots of reciprocal of quantum yield vs. olefin concentration are linear with slopes and intercepts as shown in Table 5. For photoaddition of 3-ethoxyisoindolenone to tetramethylethylene and *cis*-2-butene, the product ratios (*56:57:58* and *59:60:61:62*) are independent of olefin concentra-

Table 5. Slopes and intercepts of plots of reciprocal of quantum yield vs. olefin concentration

| Olefin | Conc. range mole/l | Slope l/mole | Intercept |
|---|---|---|---|
| Tetramethylethylene | 0.064–2.00 | 0.80 ± 0.03 | 1.63 ± 0.02 |
| 1,1-Dimethoxyethene | 0.060–2.00 | 0.32 ± 0.03 | 1.36 ± 0.03 |

tion. The observed concentration effect on quantum yield is consistent with olefin quenching of the singlet state and reacting with the triplet state.

3,3,4,4-Tetramethyl-1,2-diazetine 1,2-dioxide, biacetyl, *cis*-piperylene, and di-*t*-butyl nitroxide have been examined as quenchers for the photoaddition reactions of *50* to tetramethylethylene. Diazetine dioxide, biacetyl, and *cis*-piperylene quench formation of all three products equally and give linear Stern-Volmer plots with slopes ranging from 76 to 132 l/mole as shown in Table 6. The slope of the Stern-Volmer plot for quenching by *cis*-piperylene gives a triplet lifetime of $1.6 \times 10^{-8}$ sec for *50* in the presence of 0.125 M tetramethylethylene. As shown in Fig. 2, the Stern-Volmer plots for quenching product formation from photoaddition of 50 to tetramethylethylene by di-*t*-butyl nitroxide (DTBN) are not coincident at higher DTBN concentrations, and at higher DTBN concentrations the plots for quenching formation of ene product *57* and azepinone *58* are non-linear. DTBN quenching of photoaddition of *50* to *cis*-2-butene has also been studied. The Stern-Volmer plots for quenching formation of cycloadducts (*59* and *60*) and ene products (*61* and *62*) are linear but not coincident and have exceptionally high slopes of 2880 and 3790 l/mole, respectively (Table 6). Cycloadducts *59* and *60* are quenched equally within experimental error, whereas the ene products *61* and *62* are quenched differentially. At 0.005 M DTBN the ene product ratio is 2.4 compared with an ene product ratio of 3.1 in the absence of the quencher (Table 3). *cis*-2-Butene was not isomerized to *trans*-2-butene in the presence of DTBN to any significant extent during the course of the

Table 6. Stern-Volmer slopes for quenching photoaddition of *50* to olefins

| Olefin | Quencher | Products | Slope $M^{-1}$ |
|---|---|---|---|
| Tetramethyl-ethylene | Diazetine dioxide | *56, 57, 58* | 88 ± 5 |
| Tetramethyl-ethylene | Biacetyl | *56, 57, 58* | 76 ± 2 |
| Tetramethyl-ethylene | *cis*-Piperylene | *56, 57, 58* | 132 ± 2 |
| Tetramethyl-ethylene | DTBN | *56* | 97.5 ± 0.5 |
| cis-2-Butene | DTBN | *59, 60* | 2880 ± 35 |
| cis-2-Butene | DTBN | *61 + 62* | 3790 ± 15 |

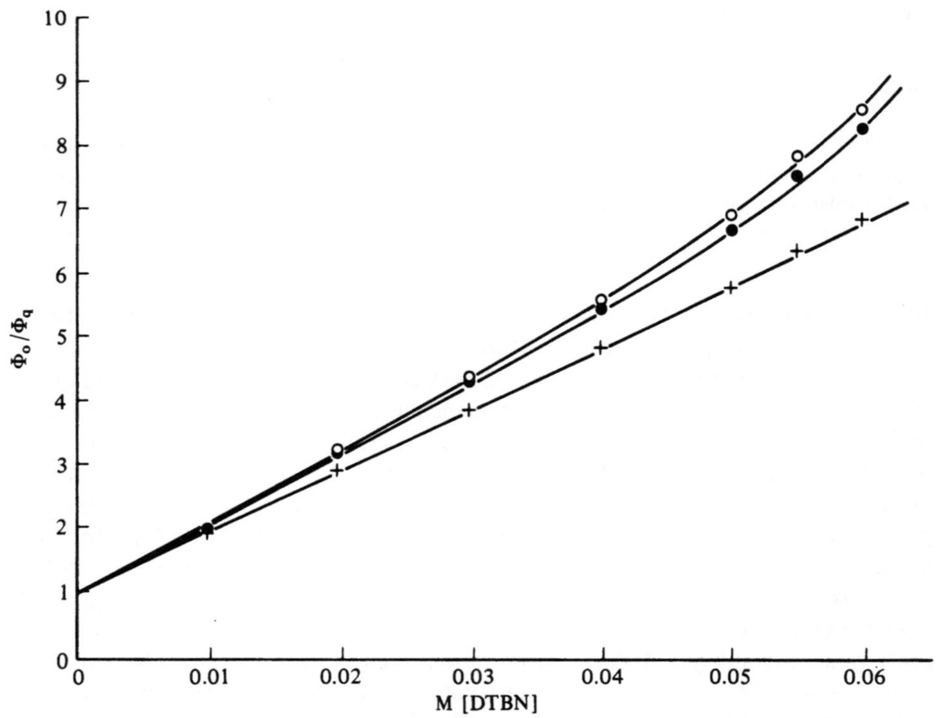


Fig. 2. Stern-Volmer plot for the quenching of (+) cycloadduct *56*, (•) ene product *57*, and (O) azepinone *58* formation from photoaddition of 3-ethoxyisoindolenone (*50*) to tetramethylethylene by di-*t*-butyl nitroxide (DTBN)

reaction. When di-*t*-butyl nitroxide is employed as a quencher of photoaddition of *50* to *cis*-2-butene and especially to tetramethylethylene some of the DTBN is destroyed. Products from the destruction of DTBN such as *70* and *71* have been isolated and characterized and result from scavenging by DTBN of radicals produced by hydrogen atom abstraction from the olefins by excited 3-ethoxyisoindolenone. The hydrogen atom abstraction process has been shown to occur in parallel with the photoaddition reactions and not to result in formation of any 3-ethoxyisoindolenone photoadducts[59].

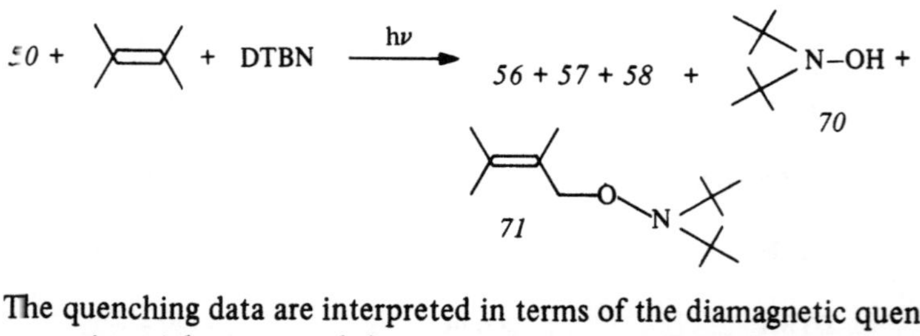


The quenching data are interpreted in terms of the diamagnetic quenchers quenching a reactive triplet state and the paramagnetic quencher quenching the reactive triplet state and/or additional intermediates (*vide infra*).

3-Ethoxyisoindolenone (*50*) sensitized *cis-trans* isomerization of *cis*-piperylene has been used in a triplet counting experiment to measure the intersystem crossing efficiency. A linear triplet counting plot is obtained which gives an intersystem crossing efficiency of 0.96 ± 0.01 and a triplet lifetime of $1.5 \times 10^{-7}$ sec. for *50* in methylene chloride solvent at ambient temperature in the absence of olefin.

## 7. Quantum Yield Measurements of Photocycloaddition of 2-(*p*-Methoxyphenyl)-2-oxazolin-4-one

Similar quantum yield results have been observed for the photocycloaddition reaction of 2-(*p*-methoxyphenyl)-2-oxazolin-4-one (*65c*) to 1,1-dimethoxyethene in benzene solvent[60]. Over the concentration range 0.24–3.00 M 1,1-dimethoxyethene, the quantum yield of product formation is inversely related to the olefin concentration. The maximum quantum yield observed is 0.027 at 0.24 M olefin. A plot of reciprocal of quantum yield vs. olefin concentration is linear with a slope of 0.75 ± 0.06 l/mole and an intercept of 37.2 ± 0.1. This result is again consistent with olefin quenching of the singlet state and reacting with the triplet state. The photoreaction is quenched by *cis*-piperylene and di-t-butyl nitroxide and linear Stern-Volmer plots are obtained. The slope of the Stern-Volmer plot with *cis*-piperylene gives a triplet lifetime of $2 \times 10^{-8}$ sec. in the presence of 0.25 M 1,1-dimethoxyethene. The linearity of the plots and the similarity of their slopes indicate that both the diamagnetic quencher and the paramagnetic quencher are quenching only the reactive triplet state. A triplet counting plot for oxazolinone sensitized *cis-trans* isomerization of piperylene gives an intersystem crossing efficiency of only 0.11. *p*-Methoxyphenyloxazolinone (*65c*) then intersystem crosses by an order of magnitude less efficiently than 3-ethoxyisoindolenone (*50*), and hence, intersystem crossing is a major cause of inefficiency in this photocycloaddition reaction.

## 8. Mechanistic Discussion

A suitable mechanism for the photoaddition reactions to olefins must account for
(1) quenching by the olefins (Table 5),
(2) the stereochemistry of the additions (Table 3),
(3) quenching by diamagnetic quenchers (Table 4),
(4) the triplet counting experiments,
(5) quenching by di-*t*-butyl nitroxide (Table 6 and Fig. 2),
(6) destruction of DTBN and
(7) the substituent effects.

A mechanism sufficient to explain these observations is complex. Therefore, a proposed mechanism will be described without discussion of alternatives, and the reader is encouraged to consult the original references for such discussion.

The fact that plots of reciprocal of quantum yield of photoaddition vs. olefin concentration are linear with positive slope (Table 5) suggests that olefin quenches the singlet state and that all triplets formed are captured by olefin in the olefin concentration range examined in competition with unimolecular decay. Inefficiency in the reaction of triplets with olefin would impart upward curvature to the plot of $\phi^{-1}$ vs olefin concentration in the low olefin concentration region. The reduction of the triplet lifetime of *50* from $1.5 \times 10^{-7}$ sec. in the absence of olefin to $1.6 \times 10^{-8}$ sec. in the presence of 0.125 M tetramethylethylene is also consistent with efficient olefin capture of the triplet state in competition with unimolecular decay. The intercepts of the plots of $\phi^{-1}$ vs. olefin concentration for photoaddition of 3-ethoxyisoindolenone (*50*) to 1,1-dimethoxyethene and tetramethylethylene are not equal (Table 5) and are larger than the reciprocal of the intersystem crossing efficiency ($\phi_{isc}{}^{-1} = 1.04$). These observations, together with the observation that the maximum quantum yield for all reactions examined is significantly less than the intersystem crossing efficiency, suggests that there is one or more energy-wasting processes after interception of the triplet state.

The lack of stereospecificity in the cycloaddition reaction and the partial stereospecificity in the ene reaction (Table 3) indicate that at least a portion of the ene reaction occurs *via* a reaction pathway different from the pathway leading to cycloadducts. A probable intermediate in the pathway to cycloadducts is a long-lived diradical such as *63* which randomizes olefin stereochemistry. The long-lived diradical may also be an intermediate in the stereorandom component of the ene reaction.

Quenching of the photoaddition reactions of 3-ethoxyisoindolenone (*50*) and *p*-methoxyphenyloxazolinone (*65c*) by diamagnetic quenchers most likely occurs by triplet energy transfer from the respective triplet states. The Stern-Volmer slopes are all of approximately the same magnitude and no differential quenching of products is observed. Diazetine dioxide is reported by Ullman and Singh not to be an effective singlet quencher[61)].

Differential quenching of photoaddition of 3-ethoxyisoindolenone to tetramethylethylene and *cis*-2-butene by di-*t*-butyl nitroxide is consistent with quenching of intermediates of different lifetimes or at different rates. The Stern-Volmer plots (Fig. 2) for DTBN quenching of the photoaddition of *50* to tetramethylethylene suggest that DTBN is quenching the triplet state of *50* as well as a subsequent intermediate leading to the ene product *57* and the azepinone *58* but not to the cycloadduct (*56*). The triplet state is implicated as one of the intermediates quenched from the similarity of the slopes of the Stern-Volmer plots with DTBN and diamagnetic quenchers (Table 6). Although DTBN is known to quench singlet states very efficiently[62)], quenching of the singlet state of *50* is probably not competitive with intersystem crossing[59)].

The Stern-Volmer plot for DTBN quenching of photoaddition to *cis*-2-butene (Table 6) is consistent with the quenching of longer lived parallel intermediates, not the triplet state of *50*, leading to ene products and cycloadducts, respectively. Quenching of a common intermediate such as the triplet state of *50*, alone or together with one or more additional intermediates, would not give the linear, noncoincident Stern-Volmer plots.

The observation that DTBN is partially destroyed with formation of di-*t*-butylhydroxylamine and di-t-butylhydroxylamine ethers when employed as a quencher of the photoaddition reactions of *50* implicates DTBN as a free radical scavenger[63)]. The free radical scavenging is thought to occur in parallel with quenching and not to be related to the quenching because the efficiency of quenching by DTBN is not related to the efficiency of its destruction. DTBN quenches photoaddition of *50* to *cis*-2-butene approximately 30 times more efficiently than addition of *50* to tetramethylethylene; however, four times more DTBN is destroyed with tetramethylethylene as the olefin than with *cis*-2-butene as the olefin. In fact with *cis*-2-butene as the olefin, the destruction of DTBN is almost insignificant.

The di-*t*-butylhydroxylamine and di-*t*-butylhydroxylamine ethers probably result from DTBN scavenging of radicals produced by hydrogen atom abstraction from the olefins by excited 3-ethoxyisoindolenone (*50*). The observed destruction of DTBN as a function of olefin structure is consistent with this mechanism. Based upon allyl radical stability and the statistical factor, excited 3-ethoxyisoindolenone should abstract hydrogen atoms more rapidly from tetramethylethylene than from *cis*-2-butene.

A kinetic mechanism consistent with the data and the previous discussion is shown in Scheme 5. The mechanism is for photoaddition of 3-ethoxyisoindolenone to tetramethylethylene and *cis*-2-butene. The mechanism is also suitable for photo-

| | | | |
|---|---|---|---|
| (1) | Ind + h$\nu$ | → | $^1$Ind* |
| (2) | $^1$Ind* + O | → | Ind + O |
| (3) | $^1$Ind* | → | $^3$Ind* |
| (4) | $^3$Ind* + O | → | $E_1$ (exciplex) |
| (5) | $^3$Ind* + O | → | $E_2$ (exciplex) |
| (6) | $^3$Ind* + O | → | Ind-H$^\bullet$ + O$^\bullet$ |
| (7) | $^3$Ind* + Q | → | Ind + (Q or $^3$Q*) |
| (8) | $E_1$ | → | ene product |
| (9) | $E_1$ | → | azepinone |
| (10) | $E_1$ | → | Ind + O |
| (11) | $E_1$ + DTBN | → | Ind + O + DTBN |
| (12) | $E_2$ | → | D (diradical) |
| (13) | $E_2$ | → | Ind + O |
| (14) | $E_2$ + DTBN | → | Ind + O + DTBN |
| (15) | D | → | ene product |
| (16) | D | → | cycloadducts |
| (17) | Ind-H$^\bullet$ + DTBN | → | Ind + di-*t*-butylhydroxylamine (73) |
| (18) | O$^\bullet$ + DTBN | → | di-*t*-butylhydroxylamine ether (21) |

Ind ≡ 3-ethoxyisoindolenone
Q ≡ quencher
DTBN ≡ di-*t*-butyl nitroxide

Scheme 5

cycloaddition of 3-ethoxyisoindolenone and 2-aryloxazolinones to 1,1-dimethoxyethene although with some modifications. Two DTBN quenchable, non-equilibrating triplet exciplexes are included as intermediates. One exciplex ($E_1$) leads to ene products with at least partial stereospecificity and to azepinone, and the other ($E_2$) leads to a diradical which subsequently gives cycloadducts and ene products stereorandomly. Quenching of the isoindolenone triplet state (step 7) is significant with biacetyl, *cis*-piperylene, and DTBN as quenchers of photoaddition to tetramethylethylene. Quenching of $E_1$ by DTBN (step 11) occurs during the photoaddition to tetramethylethylene at higher DTBN concentrations and during the photoaddition to *cis*-2-butene in the presence of DTBN. Quenching of $E_2$ by DTBN (step 14) only occurs during the photoaddition to *cis*-2-butene. Steps 6, 17 and 18, involved in the destruction of DTBN, are primarily significant in the photoreaction of *50* with tetramethylethylene. For the photoaddition of *p*-methoxyphenyloxazolinone (*65c*) to 1,1-dimethoxyethene, DTBN and *cis*-piperylene quench the reactive triplet state and no exciplex quenching occurs. Exciplex quenching by DTBN is probably a function of exciplex lifetime, and there must be a significant variation in exciplex lifetime as a function of olefin and keto imino ether structure.

Possible structures proposed for two nonequilibrating, triplet, exciplex intermediates are *n*- and *π*-type exciplexes[64]. An *n*-type exciplex utilizing the nitrogen nonbonding orbital could react stereospecifically in an ene reaction. The degree of stereospecificity would depend on the reactivity of the exciplex as a function of conformation. If the exciplex reacted exclusively when the olefin *π* system was coplanar with the *σ* framework of *50* as shown in Scheme 6, *trans*-2-butene would give *61* and *cis*-2-butene would give *62*.

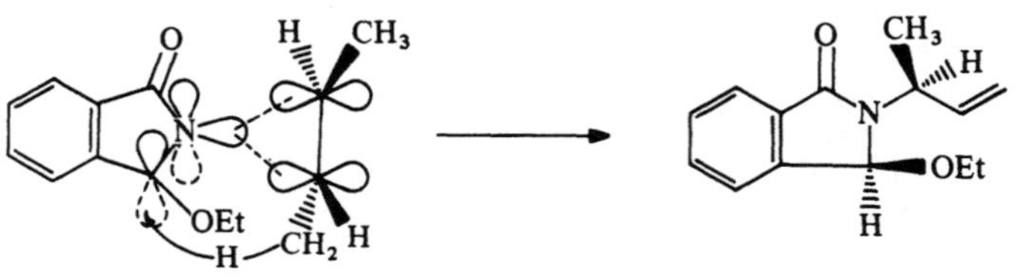


*72a* *61*

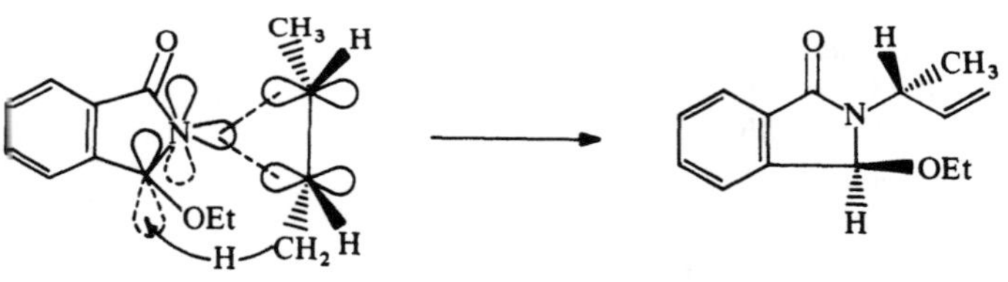


*72b* *62*

Scheme 6. A Stereospecific ene reaction from an *n*-type exciplex with the olefin *π*-structure coplaner with the *σ*-framework of 3-ethoxyisoindolenone

Analogous reactivity of an *n*-type exciplex from a conformation in which the olefin *π*-system was orthogonal to the *σ* framework of *50* would give the opposite stereochemistry. If both conformations were reactive, but unequally, partial stereospecificity would be observed. The point of intersystem crossing of an *n*-type exciplex would also be critical to the stereospecificity of the ene reaction, and only a short-lived diradical would be a possible additional intermediate. The azepinone (*58*) observed only with tetramethylethylene as the alkene could also arise from the *n*-type exciplex by participation of the carbonyl-carbon nitrogen *σ* bond. The steric effect of four methyl groups might enhance the reaction of the exciplex leading to azepinone relative to the reaction leading to ene product.

The second exciplex ($E_2$) might be a $\pi$-type exciplex such as *73* which could collapse to along-lived diradical *63* ultimately giving cycloadducts and ene products stereorandomly.

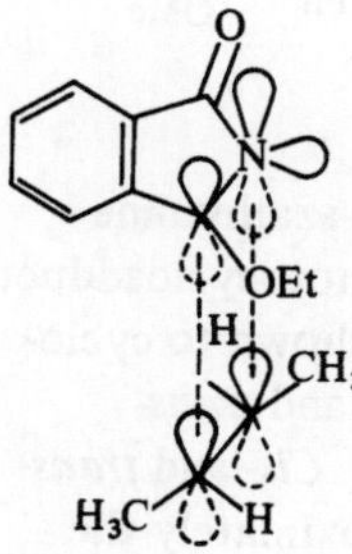

*73*

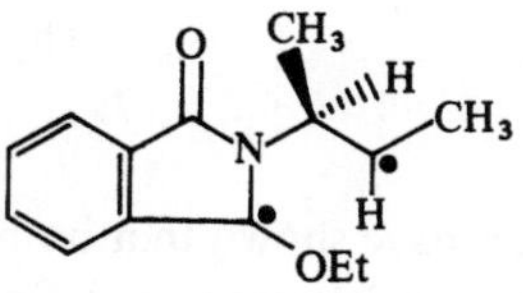

*63*

Participation of an *n*-type exciplex such as *72a* or *72b* in the photochemistry of *50* implicates a partially vacant nonbonding orbital on nitrogen in the reactive excited state. The phosphorescence lifetime of *50* points to a $\pi,\pi^*$ configuration for the lowest energy triplet state, and the substituent effects on the photoreactivity of the aryloxazolinones (*64*) implicate a $\pi,\pi^*$ configuration for the triplet state reactive in (2 + 2)-cycloaddition. Vibronically coupled[65] or equilibrating $n,\pi^*$ and $\pi,\pi^*$ triplet states of *50* might lead to both *n*- and $\pi$-type exciplexes.

The lack of reactivity of the aryloxazolinones (*65*) in photocycloaddition to many of the olefins other than 1,1-dimethoxyethene and furan probably results from efficient decay of $E_2$ or D. Exciplex $E_2$ and diradical D are proposed as intermediates in these cases for several reasons. Exciplex formation is most likely dependent on olefin ionization potential, and the ionization potential of many of the unreactive olefins are intermediate between the ionization potential of furan and 1,1-dimethoxyethene as determined from the maxima of tetracynoethylene olefin charge transfer bands[60, 66, 67]. Although *cis*-2-butene does not form a cycloadduct with 2-phenyl-2-oxazolin-4-one (*65a*), *cis*-2-butene is isomerized to *trans*-2-butene during the irradiation[52]. *Cis-trans* isomerization is expected from decay of a triplet diradical. Decay of the exciplex and diradical intermediates in competition with reaction presumably results from steric hindrance from the aryl substituent. The olefins which give cycloadducts, furan and 1,1-dimethoxyethene, are expected to produce low steric hindrance with the aryl substituent in an exciplex or diradical.

## 9. Other (2 + 2)-Photocycloadditions to Carbon-Nitrogen Double Bonds

We have observed that 6-methyl-2-phenyl-1,3-oxazin-4-one (*74*) photocycloadds regiospecifically to 1,1-dimethoxyethene[68] and to furan. The regiochemistry is the same as observed with 3-ethoxyisoindolenone (*50*) and 2-aryl-2-oxazolinone (*65*). The oxazinone (*74*) like the aryloxazolinones is not reactive with other olefins. The cycloadduct to 1,1-dimethoxyethene (*75*) is thermally labile, cleaving to 3,3-dimethoxy-2-phenyl-1-azetine (*76*) at elevated temperature.

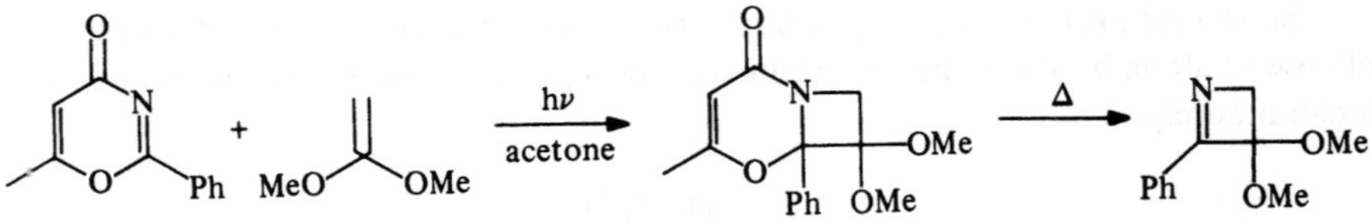

74 75 76

Swenton and co-workers have shown that both 6-azauracil and 6-azathymine photocycloadd with acetone sensitization to olefins to give (2 + 2)-photocycloadducts in high yield and with high regiospecificity[69]. 6-Azauracil has been shown to cycloadd to ethylene, tetramethylethylene, isobutylene, cyclohexene, *cis-* and *trans-*cyclooctene, ethyl vinyl ether, vinyl acetate, and isopropenyl acetate. *Cis-* and *trans-*cyclooctene give the same epimeric mixture of cycloadducts in approximately the same ratio.

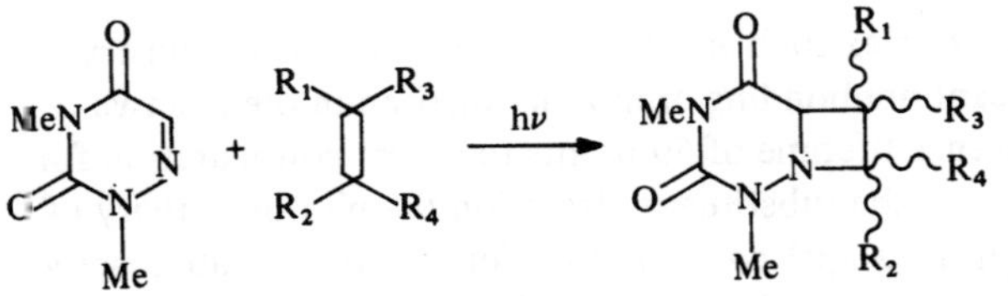

The regiospecificity of the cycloadditions is the same as the regiospecificity of the additions to *50, 65,* and *74* with respect to the carbonyl but opposite with respect to the nitrogen of the carbon-nitrogen double bond. The vinylacetate cycloadduct of 6-azauracil has been shown to be as useful intermediate for the synthesis of 5-substituted uracils[70].

Tsuge and co-workers have shown that 2,5-diphenyl-1,3,4-oxadiazole (*8*) photocycloadds in low yield regiospecifically to furan with and without benzophenone sensitization[71]. The reaction is also quenched by piperylene. The regiospecificity is analogous to that observed with *50, 65,* and *74.*

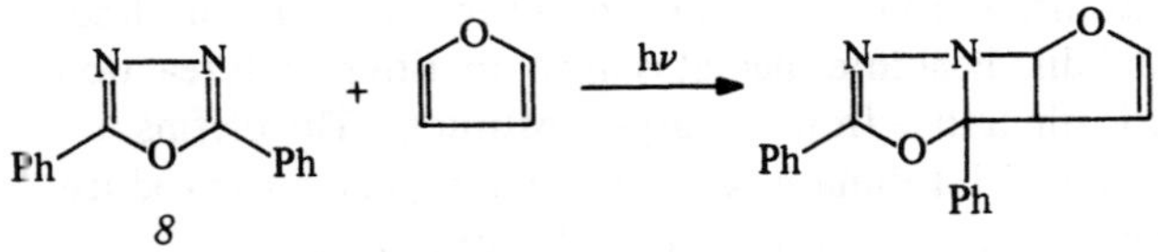

8

In comparing the various (2 + 2)-photocycloaddition reactions to carbon-nitrogen double bonds a few generalizations can be made. The reactions are uniformily triplet reactions. The configuration of the reactive excited state is probably of $\pi,\pi^*$ configuration; the cycloadditions occur regiospecifically; exciplexes and 1,4-diradicals are probably intermediates; and reactive chromophores contain conjugating electron withdrawing functional groups bonded directly to the nitrogen of the carbon-nitrogen double bond.

Other analogous systems which we have explored for reactivity in the (2 + 2)-photocycloaddition reaction include the following, none of which are reactive[72-74]:

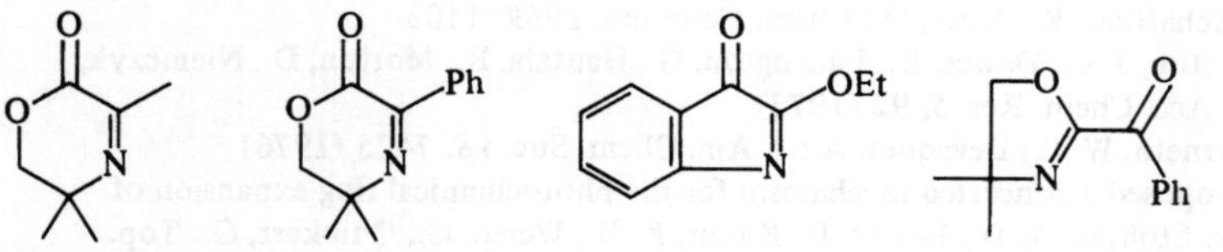


These systems do not have electron withdrawing groups bonded directly to the nitrogen of the carbon nitrogen double bond.

*Acknowledgement.* Financial support from Research Corporation, the Petroleum Research Fund, the General Medical Institute of the National Institutes of Health, and the Biomedical Research Support Grant Program of the National Institutes of Health is gratefully acknowledged. D.R.A. and G.C.C. would like to thank the University of Colorado for graduate fellowships and T.H.K., the University of Colorado Council on Research and Creative work for a Faculty Fellowship.

## IV. References

1) Chapman, O. L., Lenz, G.: Org. Photochem. *1,* 283 (1967); Eaton, P. E.: Acc. Chem. Res. *1,* 50 (1968); De Mayo, P.: Acc. Chem. Res., *4,* 41 (1971); Bauslaugh, P. G.: Synthesis *2,* 287 (1970)
2) Arnold, D. R.: Adv. Photochem. *6,* 301 (1968); Turro, N. J., Dalton, J. C., Dawes, K., Farrington, G., Hautala, R., Morton, D., Niemczyk, M., Schore, N.: Acc. Chem. Res. *5,* 92 (1972)
3) Corey, E. J., Nozoe, S.: J. Am. Chem. Soc. *86,* 1652 (1964); Dilling, W. L.: Photochem. and Photobiol. *25,* 605 (1977)
4) Searles, S. Jr., Clasen, R. A.: Tetrahedron Lett. *1965,* 1627
5) Ohta, H., Tokumaru, K.: Bull. Chem. Soc. Japan *48,* 1669 (1975)
6) Kan, R. O., Furey, R. L.: J. Am. Chem. Soc. *90,* 1666 (1968)
7) Padwa, A., Bergmark, W., Pashayan, D.: J. Am. Chem. Soc. *90,* 4458 (1968); *91,* 2653 (1969)
8) Tsuge, O., Tashiro, M., Oe, K.: Tetrahedron Lett. *1968,* 3971
9) Oe, K., Tashiro, M., Tsuge, O.: J. Org. Chem. *42,* 1496 (1977)
10) Eaton, P. E.: J. Am. Chem. Soc. *84,* 2454 (1962)
11) Corey, E. J., Bass, J. D., LeMahieu, R., Mitra, R. B.: J. Am. Chem. Soc. *86,* 5570 (1964)
12) de Mayo, P., Pete, J.-P., Tchir, M.: Can. J. Chem. *46,* 2535 (1968)
13) Padwa, A., Albrecht, F.: J. Am. Chem. Soc. *96,* 4849 (1974)
14) Koch, T. H., Sluski, R. J.: Tetrahedron Lett. *1970,* 2391
15) Comstock, W. J., Wheeler, H. L.: Am. Chem. J. *13,* 522 (1891)
16) Crockett, G. C., Koch, T. H.: J. Org. Chem. *42,* 2721 (1977)
17) Crockett, G. C., Koch, T. H.: J. Org. Chem. *42,* 2721
18) Koch, T. H., Sluski, R. J., Moseley, R. H.: J. Am. Chem. Soc. *95,* 3957 (1973)
19) Sluski, R. J.: Masters Thesis, University of Colorado 1971
20) Burns, J. M., Ashley, M. E., Crockett, G. C., Koch, T. H.: J. Am. Chem. Soc. *99,* 6924 (1977)
21) Haltiwanger, R. C., Burns, J. M., Crockett, G. C., Koch, T. H.: J. Am. Chem. Soc., submitted (1977)
22) Al-Jallo, H. N. A., Waight, E. S.: J. Chem. Soc. B *1966,* 73
23) Herz, W., Iyer, V. S., Nair, M. G., Saltiel, J.: J. Am. Chem. Soc. *99,* 2704 (1977)
24) The computer program used for these calculations is a modified version of the original program as described by R. Hoffmann: J. Chem. Phys. *39,* 1397 (1963). For a description of the modified program see S. S. Zumdahl: Ph. D. Thesis, University of Illinois 1968
25) Lamola, A. A., Hammond, G. S.: J. Chem. Phys. *43,* 2129 (1965)
26) Woodward, R. B., Hoffmann, R.: Angew. Chem., Int. Ed. Engl. *8,* 781 (1969)
27) Dewar, M. J. S.: Angew., Chem., Int. Ed. Engl. *10,* 761 (1971); Zimmerman, H. E.: Acc. Chem. Res. *4,* 272 (1971)

28) Baggiolini, E., Schaffner, K., Jeger, O.: Chem. Commun. *1969*, 1103

29) Turro, N. J., Dalton, J. C., Dawes, K., Farrington, G., Hautala, R., Morton, D., Niemczyk, M., Schore, N.: Acc. Chem. Res. *5*, 92 (1972)

30) Turro, N. J., Farneth, W. E., Devaquet, A.: J. Am. Chem. Soc. *98*, 7425 (1976)

31) Quinkert has proposed a concerted mechanism for the photochemical ring expansion of cyclobutanones: Stohrer, W. D., Jacobs, P., Kaiser, K. H., Weich, G., Quinkert, G.: Top. Curr. Chem. *46*, 181 (1974)

32) Salem, L.: J. Am. Chem. Soc., 96, 3486 (1974); Dauben, W. G., Salem, L., Turro, N. J.: Acc. Chem. Res. *8*, 41 (1975)

33) Hoffmann, R.: J. Am. Chem. Soc. *90*, 1475 (1968)

34) Stephenson, L. M., Brauman, J. I.: J. Am. Chem. Soc. *93*, 1988 (1971); Stephenson, L. M., Gibson, T. A., J. Am. Chem. Soc. *96*, 5624 (1974)

35) Lewis, F. D., Hilliard, T. A.: J. Am. Chem. Soc. *94*, 3852 (1972)

36) Agosta, W. C., Smith, A. B., III, Kende, A. S., Eilerman, R. G., Benham, J. Tetrahedron Lett. *1969*, 4517; Agosta, W. C., Smith, A. B., III: J. Am. Chem. Soc. *93*, 5513 (1971)

37) East, D. S. R., McMurray, T. B. H., Ratuakar, R.: J. Chem. Soc. Perkin I *1976*, 433

38) Padwa, A., Ku, A., Sato, E.: Tetrahedron Lett. *1976*, 2409

39) Herz, W., Iyer, V. S.: J. Org. Chem. *42*, 1573 (1977)

40) Chapman, O. L., Kane, M., Lassila, J. D., Loeschen, R. L., Wright, H. E.: J. Am. Chem. Soc. *91*, 6856 (1969)

41) Kende, A. S., Goldschmidt, Z., Izzo, P. I.: J. Am. Chem. Soc. *91*, 6858 (1969)

42) Quinkert, G.: Pure Appl. Chem. *33*, 285 (1973)

43) Sasaki, T., Eguchi, S., Ohno, M.: J. Am. Chem. Soc. *92*, 3192 (1970)

44) Bonneau, R., Joussot-Dubien, J., Salem, L., Yarwood, A. J.: J. Am. Chem. Soc. *98*, 4329 (1976); Bonneau, R., deViolet, P. F.: Compt. Rend. Acad. Sc. Paris Series C *284*, 631 (1977)

45) Lindberg, P., Bergman, R., Wickberg, B.: J. Chem. Soc. Chem. Commun. *1975*, 946

46) For a discussion of cyclopropanone chemistry see Wassermann, H. H., Clark, G. M., Turley, P. C.: Topics Curr. Chem. *47*, 73 (1974); Tilborg van, W. J. M., Steinberg, H., deBoer, T. H.: Rec. Trav. Chim. Pays-Bas *93*, 294 (1974); Tilborg van, W. J. M.,: Thesis, University of Amsterdam 1971; Turro, N. J.: Acc. Chem. Res. *2*, 25 (1969)

47) Wasserman, H. H., Baird, M. S.: Tetrahedron Lett. *1971*, 3721

48) Wasserman, H. H., Hearn, M. H., Haveaux, B., Thyes, M.: J. Org. Chem. *41*, 153 (1976)

49) Koch, T. H., Howard, K. A.: Tetrahedron Lett. *1972*, 4035

50) Howard, K. A., Koch, T. H.: J. Am. Chem. Soc. *97*, 7288

51) Koch, T. H., Rodehorst, R. M.: Tetrahedron Lett. *1972*, 4039

52) Rodehorst, R. M., Koch, T. H.: J. Am. Chem. Soc. *97*, 7298 (1975)

53) Gordon, A. J.: 14th Annual Report of Research of the Petroleum Research Fund, 1969, p. 185

54) Rawlings, D., Keute, J. S., Koch, T. H.: unpublished work

55) Beckett, A., Porter, G.: Trans. Faraday Soc. *59*, 2051 (1963); Porter, G., Suppan, P.: Trans Faraday Soc. *61*, 1664 (1965); Porter, G., Suppan, P.: Trans. Faraday Soc. *62*, 3375 (1966)

56) Baum, E. J., Wan, J. K. S., Pitts, Jr., J. N.: J. Am. Chem. Soc. *88*, 2652 (1966); Pitts, Jr., J. N., Burley, D. R., Mani, J. C., Broadbent, A. D.: J. Am. Chem. Soc. *90*, 5902 (1968)

57) Yang, N. C., McClure, D. S., Murov, S. L., Houser, J. J., Dusenbery, R.: J. Am. Chem. Soc. *89*, 5466(1967); Yang, N. C., Dusenbery, R.: J. Am. Chem. Soc. *90*, 5899 (1968); Yang, N. C., Dusenbery, R.: Mol. Photochem. *1*, 159 (1969)

58) Wagner, P. J., Capen, G.: Mol. Photochem. *1*, 173 (1969); Wagner, P. J., Kamppainen, A. E., Schott, H. N.: J. Am. Chem. Soc. *95*, 5604 (1973)

59) Anderson, D. R., Keute, J. S., Koch, T. H., Moseley, R. H.: J. Am. Chem. Soc. *99*, 6332 (1977)

60) Keute, J. S., Koch, T. H.: unpublished work

61) Ullman, E. T., Singh, P.: J. Am. Chem. Soc. *94*, 5077 (1972)

62) Weiss, D. S.: J. Photochem. *6*, 301 (1977)

63) Schuster, D. I., Barile, G. C., Liu, K.: J. Am. Chem. Soc. *97*, 4441 (1975)

64) Turro, N. J., Dalton, J. C., Dawes, K. Farrington, G., Hautala, R., Morton, D., Niemczyk, M., Schore, N.: Acc. Chem. Res. *5*, 92 (1972); Wolf, M. W., Legg, K. D., Brown, R. E., Singer, L. A., Parks, J. H.: J. Am. Chem. Soc. *97*, 4490 (1975)

65) Hochstrasser, R. M., Marzzacco, C. A. in Molecular Luminescence Lim, E. C. ed., New York: Benjamin, W. A.: 1969, p. 631

66) Arimoto, T., Osugi, J.: Chem. Lett. *1974,* 271; Arimoto, T., Osugi, J.: Rev. Phys. Chem. Japan *44,* 25 (1974)

67) Lewis, F. D., Hirsch, R. H., Roach, P. M., Johnson, D. E.: J. Am. Chem. Soc. *98,* 8438 (1976); Schore, N. E., Turro, N. J., J. Am. Chem. Soc. *97,* 2482 (1975)

68) Koch, T. H., Higgins, R. H., Schuster, H. F.: Tetrahedron Lett. *1977,* 431

69) Hyatt, J. H., Swenton, J. S.: Chem. Comm. *1972,* 1144; Swenton, J. S., Hyatt, J. H.: J. Am. Chem. Soc. *96,* 4879 (1974)

70) Swenton, J. S., Balchunis, R. J.: J. Het. Chem. *11,* 453 (1974); J. Het. Chem. *11,* 917 (1974)

71) Tsuge, O., Oe, K., Tashiro, M.: Tetrahedron *29,* 41 (1973)

72) Koch, T. H., Olesen, J. A., DeNiro, J.: J. Org. Chem. *40,* 14 (1975)

73) Koch, T. H., Olesen, J., Foy, J.: J. Org. Chem. *40,* 117 (1975)

74) Wharry, D., Koch, T. H.: unpublished work


Received January 19, 1978

# Computational Methods of Correlation Energy

**Ivan Hubač**

**Department of Mathematics, Chemical Faculty, Slovak Technical University, Bratislava, Czechoslovakia**

**Petr Čársky**

**J. Heyrovský Institute of Physical Chemistry and Electrochemistry, Czechoslovak Academy of Sciences, Prague, Czechoslovakia**

**Table of Contents**

# I. Introduction

Recent progress in *ab initio* calculations resulted in attempts to achieve so-called chemical accuracy. Since the late sixties this was reflected in the growth of the number of theoretical studies devoted to the problem of the correlation energy. Although to date a large amount of material on this topic has been accumulated, the calculations beyond the Hartree-Fock limit (other than CI calculations) are still rather exclusive. In our opinion, one of the reasons for this state is the circumstance that the developed theories are the domain of a few theoreticians, who are familiar with diagrammatic methods and who, unfortunately, are not too interested in numerical calculations oriented to chemical applications. We felt that we might contribute to the formation of a linkage between the theory and chemical applications because one of us has examined the possibilities of the perturbation theory and diagrammatic technique in its applications to excitation energies, ionization potentials and ground state energies, while the second author is interested in applications of *ab initio* calculations in the field of chemical reactivity. The aim of this review is to give a survey of existing computational methods and make them more accessible to those who want to apply them in routine calculations. Wherever possible we emphasized two practical points – the feasibility of calculations and the portion of the correlation energy involved.

Of course, reviews on correlation energy have already been published. Actually, the correlation energy was treated on several occasions in this journal[1–3]. We are not going to attempt a complete bibliography. In this section we only take note of some reviews of general importance. As regards textbooks on quantum chemistry, attention was paid to the correlation energy in the books of Parr[4] and Schaefer[5]. An attempt to show the state of the art in the seventies was made by v. Herigonte[6]. Excellent studies with the same scope were performed by van der Velde[7], who documented the discussion of various computational methods with his own model calculations, and by Kutzelnigg[8]. Other relevant papers on the electron correlation will be noted with particular topics in the following sections.

We attempted a general review of the correlation energy, although it was our aim to favor the methods based on the many-body perturbation theory and cluster expansion of the wave function. For this reason we considered it expedient to include Sections III-V in our review. These, however, should not be considered as a textbook of second quantization and many-body techniques (this has been done elsewhere[9]). We do not give a detailed physical background to these methods but discuss only features which are important for practical calculations. It is our aim to make these methods more familiar to those who are not specialists in this field. Existing computational methods of correlation energy and the relations between them are discussed in Section VI making wide use of the theoretical foundation presented in Sections III–V. Accordingly, some attention is paid to the configuration interaction method. Since there are numerous reviews on this topic (*e.g.* [10–13]) we shall note briefly only some modified versions of the CI method. The properties of the CI wave function and the correlation energy involved will, however, be discussed in connection with other methods mentioned in Section VI. In Sections VII and VIII we present some applications of the many-body perturbation theory other than for the ground state. Otherwise, this review is oriented to the ground state.

## II. The Conservation of Correlation Energy

In the field of *ab initio* calculations there exists an apparent controversy. On the one hand, the neglect of the correlation energy is demonstrated theoretically by the crude approximation of the Hartree-Fock treatment while on the other hand, the majority of calculations are performed just at the level of the Hartree-Fock approximation mostly with meaningful results. The fact that correlation effects may be disregarded in many cases has been recognized a long time ago. The tendency to cancel correlation effects in some chemical processes has been noted by several authors[14–16] in the early sixties. The rules for the conservation of correlation energy may be expressed[3] as follows:

1. The number of electron pairs must be conserved.
2. The spatial arrangement must be approximately maintained for electron pairs which are the nearest neighbours.

The conservation of correlation energy was tested systematically by Snyder and Basch[17, 18] on a large set of chemical reactions having closed shell reactants and products. The theoretical (SCF) heats of reactions were claimed to be more accurate than those obtained using semiempirical relations of bond energies for reactions of strained molecules, or those not well represented by a single valence-bond structure. However, Snyder and Basch concluded[17, 18] that if a level of chemical accuracy is to be approached, some semiquantitative prediction of the change in correlation energy is required. Obviously, in many reactions of closed shell molecules the second of the two conditions noted is not satisfied. A typical example for the violation of the second rule is the dimerization 2 $BH_3 \longrightarrow B_2H_6$. The SCF calculation[19] gives the value of 87 kJ/mole for the energy of dimerization of $BH_3$. If the correlation energy is included (by the CEPA method) one arrives at the value[19] of 153 kJ/mole. An ideal example for the conservation of correlation energy is the internal rotation in ethane. Here the two requirements are perfectly satisfied and it is therefore not surprising that the SCF calculations give excellent agreement with experiment (for a more detailed discussion see Ref.[20]). Another example, where the two rules are well satisfied, is the following reaction:

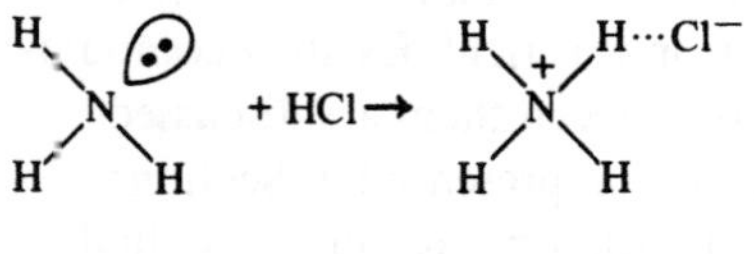


(1)

For this process it was possible to predict gaseous $NH_4Cl$ by mere SCF calculations[21]. Its existence was later established experimentally[22]. The conservation of correlation energy may also be expected in the so-called isodesmic reactions. This concept introduced by Hehre and collaborators[23] refers to reactions in which there is retention of the number of bonds of a given formal type, but with a change in their relation to one another. The degree of conservation of the spatial relation evidently depends on the particular case. The following reactions[24, 25] represent almost ideal cases:

(2)

$RO^- + CH_3OH \rightleftharpoons CH_3O^- + ROH$ (3)

$RCC^- + HCCH \rightleftharpoons HCC^- + RCCH$ (4)

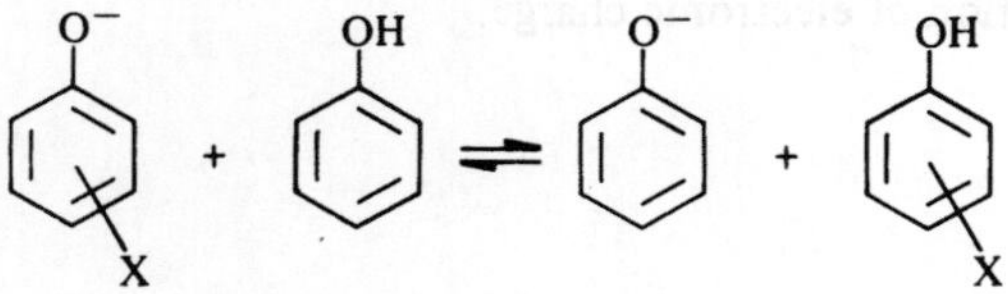
(5)

Often one can arrive at heats of reactions indirectly. For example, the reaction

$NaCl \longrightarrow Na + Cl$ (6)

does not satisfy any of the two conditions for the conservation of correlation energy. A near invariance of the correlation energy may however be assumed for the ionic decomposition

$NaCl \longrightarrow Na^+ + Cl^-$ (7)

NaCl being highly ionic, the two requirements are well satisfied. Since the correlation energies for Na, Cl, $Na^+$ and $Cl^-$ are well known, one can arrive in this way[26] at the binding energy of NaCl with near chemical accuracy. Indirectly, use is also made of the so-called "bond separation" reactions[23]. These are the isodesmic processes of the type

$X{-}Y{-}Z + Y \longrightarrow X{-}Y + Y{-}Z$ (8)

as for example

$CH_3CHO + CH_4 \longrightarrow C_2H_6 + H_2CO$ (9)

which are chemically of little interest alone but combinations of which may provide valuable results[27, 28].

As the opposite to the examples above, we now discuss processes that involve the dissociation of a bond. Here the change in correlation energy is extremely large and the Hartree-Fock approximation is inherently incapable of giving a reasonable account of the heats of reaction. The most striking case is the $F_2$ dissociation, for which the SCF calculations[29] favor two F atoms with respect to $F_2$ by 132 kJ/mole. A discussion of the role of the correlation energy in dissociation processes is given in Section VI.I.

To summarize the discussion in this section, it is fair to state that the Hartree-Fock approximation is adequate in many cases and that the importance of correlation effects is sometimes overestimated. It appears that the basis set effects may be larger than the correlation effects and that the errors in theoretical heats of formation may be mainly due to the limited basis set used[30, 31]. Very instructive discussions on this problem were reported for the inversion barrier in ammonia[6, 20, 32] and the barrier to rotation[20] in $H_2O_2$. Finally, it should be noted that the Hartree-Fock approximation is also justifiable for the one-electron properties because the latter do not depend explicitly on relative positions of electrons (electron correlation), but rather on the overall distribution of electronic charge.

## III. Second Quantization

The aim of this section is to familiarize the reader with the second quantization and the many-body diagrammatic techniques which are now widely used in up-to-date quantum chemistry. These techniques are very efficient since they permit the formulation of the problem by means of diagrams from which the explicit formula can be obtained. Another advantage is that the problem of spin can be handled very simply. This approach also permits us to have a "microscopic" view of the problem (as will be seen in the study of ionization potentials, excitation energies, interaction of two molecular systems *etc.*).

One of the most important concepts of quantum chemistry is the Slater determinant. Most quantum chemical treatments are made just over Slater determinants. Nevertheless, in many problems the formulation over Slater determinants is not very convenient and the derivation of final expressions is very complicated. The advantage of second quantization lies in the fact that this technique permits us to arrive at the same expressions in a considerably simpler way. In second quantization a Slater determinant is represented by a product of creation and annihilation operators. As will be shown below, the Hamiltonian can also be expressed by creation and annihilation operators and thus the eigenvalue problem is reduced to the manipulation of creation and annihilation operators. This manipulation can be done diagrammatically (according to certain rules which will be specified later) and from the diagrams formed one can write down the final mathematical expression. In the traditional way a Slater determinant $|\Phi\rangle$ is specified by one-electron functions as follows:

$$|\Phi\rangle = \frac{1}{(N!)^{1/2}} \sum_{\mathrm{P}} (-1)^p \, \mathrm{P} \prod_{i=1}^{N} \varphi_{A_i}(x_{A_i}) \tag{10}$$

where $\varphi$ represents spin-orbitals and $x$ represents the space and spin-coordinates. Consider now the order of one-electron functions according to increasing orbital energy. In an equivalent way we can describe the Slater determinant by specifying which $\varphi_{A_i}$ from a given set enter the Slater determinant and which do not. This may be expressed by the state vector $|n_1 n_2 \ldots n_i \ldots\rangle$, where the indices $n$ can have the value 0 or 1 depending on whether the spin-orbital $\varphi_{A_i}$ is occupied or not. The numbers $n_i$ are called occupation numbers and this form of representation of Slater determinants is called the occupation number representation. On the basis of vectors $|n_1 n_2 \ldots\rangle$ we shall define the creation ($X_i^+$) and annihilation ($X_i$) operators. The annihilation operator is defined as follows:

$$X_i |n_1 n_2 \ldots n_i \ldots\rangle = (-1)^{S_i} n_i |n_1 n_2 \ldots (n_i - 1) \ldots\rangle \tag{11}$$

where

$$S_i = \sum_{k=1}^{i-1} n_k \tag{12}$$

The quantity $(-1)^{S_i}$ gives us the correct sign of the determinant. For purposes of quantum chemistry it is more convenient to specify the state vectors in such a way that only the occupied spin-orbitals are listed in the vector. Therefore instead of $|n_1 n_2 \ldots\rangle$ we write $|A_1 A_2 \ldots\rangle$. In this case, we define the annihilation operator $X_{A_i}$ as

$$X_{A_i} |A_i A_2 \ldots A_N\rangle = |A_2 \ldots A_N\rangle \tag{13}$$

$$X_{A_i} |A_1 \ldots A_N\rangle = 0 \text{ where } i \neq 1, \ldots, N \tag{14}$$

We note that the annihilation operator is defined in such a way that it always acts on the first component of the state vector. Therefore we have to commute the state $A_i$ to the first component of state vector $|A_1 A_2 \ldots A_i \ldots\rangle$ which ensures us the correct sign. Next we define the vacuum state vector $|0\rangle$ as follows:

$$X_{A_i} |A_i\rangle = |0\rangle \tag{15}$$

The vacuum state $|0\rangle$ (so-called true physical vacuum state) is the state with no particles. Let us suppose that our vacuum state $|0\rangle$ and the spin-orbitals $|A_1\rangle$, $|A_2\rangle \ldots$ satisfy the following relations

$$\langle 0|0\rangle = 1 \tag{16}$$

and

$$\langle A_i|A_j\rangle = \delta_{ij} \tag{17}$$

The Hermitian conjugate state to $|0\rangle$ can be written as

$$\langle 0| = \langle A_i| X_{A_i}^+ \tag{18}$$

Let us multiply Eq. (18) by $|0\rangle$ which gives us

$$\langle 0|0\rangle = \langle A_i | X^+_{A_i} | 0\rangle \tag{19}$$

According to Eq. (17) we can write

$$\langle A_i | X^+_{A_i} | 0\rangle = 1 = \langle A_i | A_i\rangle \tag{20}$$

from which immediately follows that

$$X^+_{A_i} | 0\rangle = | A_i\rangle \tag{21}$$

Operator $X^+_{A_i}$ shall be called the creation operator. Thus, as in Eqs. (13) and (14) we can define

$$X^+_{A_i} | A_1 \ldots A_N\rangle = | A_i A_1 \ldots A_N\rangle \tag{22}$$

where $i \neq 1, \ldots, N$ and

$$X^+_{A_i} | \ldots A_i \ldots\rangle = 0 \tag{23}$$

In occupation number representation Eq. (23) becomes

$$X^+_i | n_1 n_2 \ldots n_i \ldots\rangle = (-1)^{S_i} (1-n_i) | n_1 n_2 \ldots (n_i + 1) \ldots\rangle \tag{24}$$

where $S_i$ is defined in the same way as in Eq. (12). From Eqs. (21) and (22) it follows that any Slater determinant can be written as a product of creation operators acting on the vacuum state $|0\rangle$. Therefore we can write

$$| A_1 \ldots A_N\rangle = \prod_{i=1}^{N} X^+_{A_i} | 0\rangle \tag{25}$$

Here we can see that instead of a determinant we have a product of creation operators.

The reason why we pay such attention to second quantization is that any one-particle P-operator and two-particle Q-operator can be written as

$$\mathrm{P} = \sum_{AB} \langle A | \mathrm{p} | B\rangle X^+_A X_B \tag{26}$$

and

$$\mathrm{Q} = \frac{1}{2} \sum_{ABCD} \langle AB | \mathrm{q} | CD\rangle X^+_A X^+_B X_D X_C \tag{27}$$

where $X^+_A$ and $X_B$ are creation and annihilation operators respectively, defined on the one-particle basis set $|A\rangle, |B\rangle, \ldots$ . Hereafter we shall use the following notation

$$|A\rangle = |a\rangle|\alpha\rangle \tag{28}$$

where $|A\rangle$ stands for spin-orbital, $|a\rangle$ stands for orbital and $|\alpha\rangle$ is the spin-function. Therefore the Hamiltonian H can be written as

$$\mathrm{H} = \sum_{AB} \langle A|z|B\rangle\, X_A^+ X_B + \frac{1}{2} \sum_{ABCD} \langle AB|v|CD\rangle\, X_A^+ X_B^+ X_D X_C \tag{29}$$

where the notation is such that $A$ and $C$ ($B$ and $D$) have the same space and spin coordinates. The sequence of the creation and annihilation operators in the second term of Eq. (29) is very important. We see that in the Hamiltonian we also have the products of creation and annihilation operators. Note that the Hamiltonian in this form does not depend on the number of particles and it is completely defined by the one-electron functions $|A\rangle, |B\rangle \ldots$ . This property will be used in Section VII to derive explicit expressions for ionization potentials, excitation energies and similar quantities.

To proceed further, we have to know how to handle the products of creation and annihilation operators. It is Wick's theorem which tells us how to deal with the products of these operators. Before presenting Wick's theorem we have to introduce some necessary definitions and relations. The creation and annihilation operators satisfy the anticommutation relation

$$[X_A^+, X_B]_+ = X_A^+ X_B + X_B X_A^+ = \langle A|B\rangle \tag{30}$$

It is this property which ensures that the $N$-particle systems obey Fermi statistics. (For a detailed understanding we recommend the textbooks[33–36].)

Next, we have to define the normal product ($n$-product) and contraction (pairing). The normal product is defined in the following way:

$$n[\mathrm{M}_1 \ldots \mathrm{M}_N] = (-1)^p \mathrm{M}_1 \ldots \mathrm{M}_N \tag{31}$$

where $\mathrm{M}_i$ stands either for creation or annihilation operator. All creation operators on the right hand side of Eq. (31) stand to the left of all annihilation operators with $p$ being the parity of the corresponding permutation.

The contraction (pairing) of two operators $\mathrm{M}_1$ and $\mathrm{M}_2$ is defined as

$$\overbrace{\mathrm{M}_1\mathrm{M}_2} = \mathrm{M}_1\mathrm{M}_2 - n[\mathrm{M}_1\mathrm{M}_2] \tag{32}$$

Owing to the anticommutation relation (30) the only non-zero contraction is

$$\overbrace{X_A X_B^+} = \langle A|B\rangle \tag{33}$$

Note that this contraction is a number and that the creation operator stands to the right of the annihilation operator. All other contractions are zero.

A normal product with contraction is given by

$$n[\mathrm{M}_1\mathrm{M}_2 \ldots \overbrace{\mathrm{M}_i \ldots \mathrm{M}_j} \ldots \overbrace{\mathrm{M}_k \ldots \mathrm{M}_l} \ldots \mathrm{M}_N] =$$

$$= (-1)^p \overbrace{\mathrm{M}_i\mathrm{M}_j}\,\overbrace{\mathrm{M}_k\mathrm{M}_l} \ldots n[\mathrm{M}_1\mathrm{M}_2 \ldots \mathrm{M}_N] \tag{34}$$

where $p$, as in definition (31), is the parity of the corresponding permutation.

Now we are ready to formulate the time independent Wick's theorem:

$$\mathrm{M}_1\mathrm{M}_2 \ldots \mathrm{M}_N = n[\mathrm{M}_1\mathrm{M}_2 \ldots \mathrm{M}_N] + n[\overbrace{\mathrm{M}_1\mathrm{M}_2} \ldots \mathrm{M}_N] + \ldots$$

$$\ldots n[\overbrace{\mathrm{M}_1\mathrm{M}_2 \ldots \mathrm{M}_N}] + n[\overbrace{\mathrm{M}_1\mathrm{M}_2}\,\overbrace{\mathrm{M}_3\mathrm{M}_4} \ldots \mathrm{M}_N] + \ldots \tag{35}$$

Expressed in words, the product of creation and annihilation operators is equal to a normal product of these operators plus the sum of normal products with one contraction plus the sum of normal products with two contractions *etc.*, up to the normal product where all operators are contracted. This theorem can be generalized (to the so-called generalized Wick's theorem) in the way that also a product of the following form can be handled:

$$n[\mathrm{M}_1 \ldots \mathrm{M}_i]\,n[\mathrm{M}_{i+1} \ldots \mathrm{M}_k] \ldots n[\ldots \mathrm{M}_l] \tag{36}$$

Its expression is similar to Eq. (35) but omits contractions between operators within the same normal product. An important consequence of Eqs. (15) and (18) should be emphasized, that

$$\langle 0 | n[\mathrm{M}_1 \ldots \mathrm{M}_N] | 0 \rangle = 0 \tag{37}$$

is valid if at least one operator in $n[\mathrm{M}_1 \ldots \mathrm{M}_N]$ is not contracted.

In Eq. (25) the number of creation operators is equal to the number of electrons. With many-electron systems the number of creation and annihilation operators is large and their handling becomes unpleasant, *e.g.* due to the large number of terms appearing upon the application of Wick's theorem. In the Slater determinant for a doubly excited configuration $|\Phi_{ab}^{rs}\rangle$ for example, the only important orbitals are $a$, $b$ (occupied in the ground state) and $r$, $s$ (unoccupied in the ground state). For practical reasons it would therefore be very convenient to describe such a determinant only with operators of pertinent orbitals. This can be done using the idea of hole-particle formalism.

The essence of the hole-particle formalism lies in a new meaning given the vacuum state. Consider now the closed shell ground state Slater determinante $|\Phi_0\rangle$ expressed by means of Eq. (25) as

$$|\Phi_0\rangle = \prod_{i=1}^{N} \mathrm{X}_{A_i}^+ |0\rangle \tag{38}$$

Let us adopt $|\Phi_0\rangle$ as a new vacuum state. We shall call it the Fermi vacuum state. With respect to the Fermi vacuum state $|\Phi_0\rangle$ we can now define new creation and annihilation operators which in contrast to the $\mathrm{X}^+$, $\mathrm{X}$ operators shall be designated

$Y^+$, Y. Let us label the occupied spin-orbitals as $|A'\rangle$ and the virtual spin-orbitals as $|A''\rangle$. The creation ($Y^+$) and annihilation (Y) operators are then defined in the following way:

$$Y^+_{A'} = X_{A'}; \quad Y_{A'} = X^+_{A'} \tag{39}$$

$$Y^+_{A''} = X^+_{A''}; \quad Y_{A''} = X_{A''} \tag{40}$$

From this definition it is evident that application of $Y^+_{A'}$ to the Fermi vacuum is equivalent to annihilation of a particle (or creation of a hole) in $|\Phi_0\rangle$. The effect of $Y_{A'}$ on the Fermi vacuum state is the creation of a particle (or annihilation of a hole) in $|\Phi_0\rangle$. The effect of $Y^+_{A''}$ on the Fermi vacuum is the creation of a particle in the virtual spin-orbitals and finally, the effect of $Y_{A''}$ is the annihilation of a particle in virtual spin-orbitals. Thus *e.g.*, a singly excited Slater determinant $|\Phi^r_a\rangle$ can be described as

$$X^+_{A''_r} X_{A'_a} |\Phi_0\rangle = Y^+_{A''_r} Y^+_{A'_a} |\Phi_0\rangle \tag{41}$$

The state $|\Phi_0\rangle$ itself evidently corresponds to the state having the number of particles and holes equal to zero.

For Y operators the same anticommutation relation can be found as for X operators, *i.e.*

$$[Y^+_{A'}, Y_{B'}]_+ = \langle A' | B' \rangle \tag{42}$$

and

$$[Y^+_{A''}, Y_{B''}]_+ = \langle A'' | B'' \rangle \tag{43}$$

For Y operators, as with X operators, we can now define a normal product (which we shall designate $N[\quad]$), a contraction, as well as a normal product with contractions just by replacing X operators in the formulas with Y operators. The same holds for Wick's theorem and the generalized Wick's theorem. Equations (42) and (43) show that we have two sets of Y operators: one operating on the holes (42) and the other on the particles (43). To arrive at a more convenient form of anticommutation relations we define the following functions

$$\tau(A') = 0; \quad \tau(A'') = 1 \tag{44}$$

$$\vartheta(A') = 1; \quad \vartheta(A'') = 0 \tag{45}$$

The contractions of Y operators may now be expressed through X operators as follows

$$\overbrace{X^+_A X^+_B} = 0; \quad \overbrace{X^+_A X_B} = \vartheta(A) \langle A | B \rangle \tag{46}$$

$$\overline{X_A X_B} = 0; \quad \overline{X_A X_B^+} = \tau(A) \langle A | B \rangle \tag{47}$$

Hereafter, we shall use the X operators in the sense of (46) and (47). It should therefore be kept in mind that we are working in the hole-particle formalism.

Wick's theorem (35) which gives us the prescription for treating a product of operators may of course be applied to the Hamiltonian, expressed in the second quantization formalism (29). This leads to the Hamiltonian in a form which is of primary importance in perturbation treatments. This form of the Hamiltonian which is called the normal product form is:

$$\begin{aligned} H = {} & \langle \Phi_0 | H | \Phi_0 \rangle + \sum_{AB} \langle A | f | B \rangle N[X_A^+ X_B] \\ & + \frac{1}{2} \sum_{ABCD} \langle AB | v | CD \rangle N[X_A^+ X_B^+ X_D X_C] \end{aligned} \tag{48}$$

where $\langle \Phi_0 | H | \Phi_0 \rangle$ is the Hartree-Fock ground state energy for a closed shell system and f is the Hartree-Fock operator.

## IV. Many-Body Rayleigh-Schrödinger Perturbation Theory (MB-RSPT)

### A. Brief Description

What is commonly referred to as the MB-RSPT is developed by second quantization and Wick's theorem which are used to give the diagrammatic description of ordinary time-independent Rayleigh-Schrödinger perturbation theory. The use of the term "many-body", which originates from nuclear physics, is justifiable because the explicit expressions in MBPT are expressed in terms of matrix elements of the spin-orbitals, which reflect the many-electron interaction, in contrast to the ordinary RSPT where the matrix elements are expressed over Slater determinants. The advantage of using Feynman-like diagrams is that they give a "microscopic" view of the electron interaction in atoms and molecules. Moreover, the diagrammatic description of individual terms of the perturbation expansion permits us to use the "linked cluster" theorem which has no analogy in the ordinary RSPT.

Here we shall follow the derivation of time-independent MB-RSPT in its main features as is described in Ref.[9]. Let us assume that a perturbed Hamiltonian of an atomic or molecular system, K, may be split as

$$K = K_0 + W \tag{49}$$

where $K_0$ is the unperturbed Hamiltonian and W is the perturbation. In order to obtain a direct expression for the correlation energy we use the notation K for our Hamiltonian. As will later be seen K differs from the Hamiltonian (29) by a scalar quantity. We assume that the following equations hold for K and $K_0$ operators

$$\mathsf{K}\,|\Psi_i\rangle = \mathsf{k}_i\,|\Psi_i\rangle \tag{50}$$

$$\mathsf{K}_0\,|\Phi_i\rangle = \kappa_i\,|\Phi_i\rangle \tag{51}$$

Furthermore we assume that the complete solution of Eq. (51) is known. Our goal is to find the solution of Eq. (50) under the assumption that $|\Phi_i\rangle$ changes into the state $|\Psi_i\rangle$ if the perturbation W is switched on. We shall not go into details of the derivation. Instead, we state that for $k_i$ the following RSPT expansion holds:

$$k_i = \kappa_i + \sum_{n=0}^{\infty} \langle\Phi_i\,|\,\mathsf{W}[\mathsf{Q}_i(\mathsf{W} + \kappa_i - k_i)]^n\,|\,\Phi_i\rangle \tag{52}$$

where

$$\mathsf{Q}_i = \sum_{\substack{j \\ (j \neq i)}} \frac{|\Phi_j\rangle\langle\Phi_j|}{\kappa_i - \kappa_j} = \frac{1 - |\Phi_i\rangle\langle\Phi_i|}{\kappa_i - \mathsf{K}_0} \tag{53}$$

Equation (52) can be solved iteratively. We can collect the terms having the same order of perturbation and therefore can write

$$k_i = \sum_{j=0}^{\infty} k_i^{(j)} \tag{54}$$

where $k_i^{(j)}$ is the $j$-th order contribution. The terms up to the third order have the following forms:

$$k_i^{(0)} = \langle\Phi_i\,|\,\mathsf{K}_0\,|\,\Phi_i\rangle \tag{55}$$

$$k_i^{(1)} = \langle\Phi_i\,|\,\mathsf{W}\,|\,\Phi_i\rangle \tag{56}$$

$$k_i^{(2)} = \langle\Phi_i\,|\,\mathsf{W}\,\mathsf{Q}_i\,\mathsf{W}\,|\,\Phi_i\rangle \tag{57}$$

$$k_i^{(3)} = \langle\Phi_i\,|\,\mathsf{W}\,\mathsf{Q}_i(\mathsf{W} - k_i^{(1)})\,\mathsf{Q}_i\,\mathsf{W}\,|\,\Phi_i\rangle \tag{58}$$

The whole problem of calculating $k_i$ (at least up to the third order) is now reduced to the calculation of individual terms (55)–(58). The second quantization formalism has the advantage that these terms can be calculated easily by making use of the diagrammatic technique which will be demonstrated in Section IV.B.

We have already shown that the Hamiltonian

$$\mathsf{H} = \mathsf{Z} + \mathsf{V} \tag{59}$$

where $\mathsf{Z} = \sum_i \mathsf{z}(i)$, $\mathsf{V} = \sum_{i<j} \mathsf{v}(i, j)$ (60)

are one-particle and two-particle operators respectively, can be written in the normal product form:

$$\mathrm{H} = \langle \Phi_0 | \mathrm{H} | \Phi_0 \rangle + \sum_{AB} \langle A | \mathrm{f} | B \rangle N[\mathrm{X}_A^+ \mathrm{X}_B] + \frac{1}{2} \sum_{ABCD} \langle AB | \mathrm{v} | CD \rangle N[\mathrm{X}_A^+ \mathrm{X}_B^+ \mathrm{X}_D \mathrm{X}_C] \tag{61}$$

where, as was already said, the quantity $\langle \Phi_0 | \mathrm{H} | \Phi_0 \rangle$ is the Hartree-Fock energy of the closed shell ground state. We assume that this operator satisfies the following Schrödinger equation

$$\mathrm{H} | \Psi_i \rangle = E_i | \Psi_i \rangle \tag{62}$$

$$\mathrm{H}_0 | \Phi_i \rangle = e_i | \Phi_i \rangle \tag{63}$$

Let us write

$$\mathrm{K} = \mathrm{H} - \langle \Phi_0 | \mathrm{H} | \Phi_0 \rangle \tag{64}$$

and

$$\mathrm{K}_0 = \mathrm{H}_0 - \langle \Phi_0 | \mathrm{H}_0 | \Phi_0 \rangle \tag{65}$$

Then, the quantity $k_i$ is exactly the correlation energy of the $i$-th electronic state. The simplest partitioning of the Hamiltonian (61), according to the form (49), is

$$\mathrm{K}_0 = \sum_{AB} \langle A | \mathrm{f} | B \rangle N[\mathrm{X}_A^+ \mathrm{X}_B] \tag{66}$$

$$\mathrm{W} = \frac{1}{2} \sum_{ABCD} \langle AB | \mathrm{v} | CD \rangle N[\mathrm{X}_A^+ \mathrm{X}_B^+ \mathrm{X}_D \mathrm{X}_C] \tag{67}$$

Next we shall assume that our spin-orbital basis $|A\rangle$, $|B\rangle, \ldots$ is the Hartree-Fock basis, *i.e.*

$$\mathrm{f} | A \rangle = \epsilon_A | A \rangle \tag{68}$$

where $\epsilon_A$ are the Hartree-Fock orbital energies. Then, $\mathrm{H}_0$ in Eqs. (63) and (65) is the Hartree-Fock operator, $e_i$ in Eq. (63) is the sum of Hartree-Fock orbital energies over occupied spin-orbitals and Eq. (66) can be written in the diagonal form

$$\mathrm{K}_0 = \sum_{A} \epsilon_A N[\mathrm{X}_A^+ \mathrm{X}_A] \tag{69}$$

Examine now, using Eqs. (64), (65) and (69), the expression for the correlation energy of the ground state, *i.e.* $i = 0$, in Eq. (54). The individual terms (55)–(58) have a very simple form in this case, namely

$$k_0^{(0)} = 0 \tag{70}$$

$$k_0^{(1)} = 0 \tag{71}$$

$$k_0^{(2)} = \langle \Phi_0 | \mathsf{W}\, \mathsf{Q}_0\, \mathsf{W} | \Phi_0 \rangle \tag{72}$$

$$k_0^{(3)} = \langle \Phi_0 | \mathsf{W}\, \mathsf{Q}_0\, \mathsf{W}\, \mathsf{Q}_0\, \mathsf{W} | \Phi_0 \rangle \tag{73}$$

Here we see that use of this partitioning of the Hamiltonian, the Møller-Plesset partitioning, *i.e.*, $\mathsf{K}_0$ is given by Eq. (69) and $\mathsf{W}$ is given by Eq. (67), is very profitable since the number of terms in the perturbation expansion (70)–(73) is considerably reduced. Of course, use can be made of some other partitioning of the Hamiltonian, *e.g.* that so-called Epstein-Nesbet partitioning, which in principle differs from the Møller-Plesset partitioning in so far as that the diagonal terms $\langle ij | \mathsf{v} | ij \rangle - \langle ij | \mathsf{v} | ji \rangle$ are shifted from $\mathsf{W}$ to $\mathsf{K}_0$.

Our problem of obtaining $k_0$ up to the third order is now reduced to calculating terms (72) and (73) which we shall execute diagrammatically in Sections IV.B and IV.C.

## B. Second Order Contribution to the Correlation Energy. An Example of the Elementary Diagrammatical Approach

### 1. Construction of the Diagrams

We use the second quantization formalism to express the second order contribution to the correlation energy of the closed-shell ground state. By substituting the expression (67) for $\mathsf{W}$ in Eq. (72) we obtain

$$k_0^{(2)} = \frac{1}{4} \sum_{ABCD} \langle AB | \mathsf{v} | CD \rangle \sum_{EFGH} \langle EF | \mathsf{v} | GH \rangle$$

$$\times \langle \Phi_0 | N[\mathsf{X}_A^+ \mathsf{X}_B^+ \mathsf{X}_D \mathsf{X}_C] \mathsf{Q}_0 N[\mathsf{X}_E^+ \mathsf{X}_F^+ \mathsf{X}_H \mathsf{X}_G] | \Phi_0 \rangle \tag{74}$$

Let us first make a remark concerning $\mathsf{Q}_0$. By using Eqs. (51), (63) and (65), we obtain the denominator of $\mathsf{Q}_0$ in Eq. (53)

$$\kappa_0 - \mathsf{K}_0 = e_0 - \mathsf{H}_0 \tag{75}$$

From the binomial expansion we obtain

$$\frac{1}{e_0 - \mathsf{H}_0} = \frac{1}{e_0} \left( 1 + \frac{\mathsf{H}_0}{e_0} + \frac{\mathsf{H}_0^2}{e_0^2} + \ldots \right) \tag{76}$$

from which we get

$$\frac{1}{e_0 - H_0} |\Phi_n\rangle = \frac{1}{e_0 - e_n} |\Phi_n\rangle \tag{77}$$

Hence, $Q_0$ is responsible only for a denominator of the final expression. The denominator will contain orbital energies because

$$e_{n(k)} - e_0 = \sum_{i=1}^{k} (\epsilon_{A_i''} - \epsilon_{A_i'}) \tag{78}$$

where $k$ is the $k$-fold excitation.

We already know that in order to calculate expression (74) using the generalized Wick's theorem, we have to perform contractions between the $N$-products on the right hand side of Eq. (74). We also know that according to Eq. (37) all the operators have to be contracted. Here the diagrams can be introduced because these contractions can be represented diagrammatically.

We shall assign the following diagram to the quantity $\langle AB|v|CD\rangle N[X_A^+ X_B^+ X_D X_C]$.

Fig. 1

The diagram in Fig. 1 is drawn in such a manner that the lines leaving the vertex (which is marked by a dot) correspond to creation operators, and the entering lines to annihilation operators. The dashed line connecting the vertices shall be called the interaction line. This line stands for the two-electron potential $1/r_{ij}$. Then we can schematically rewrite Eq. (74) in the form

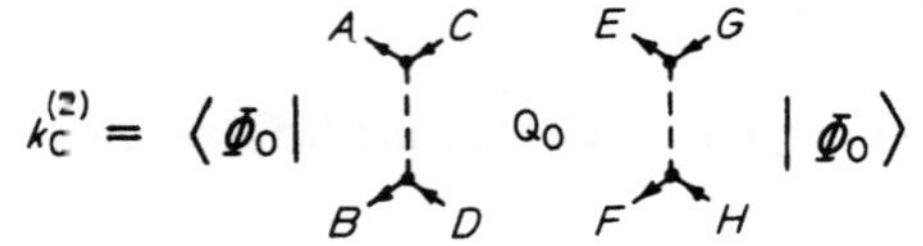

(79)

Let us omit $Q_0$ for a moment. The rules used to calculate it will be given later. Thus, we are left with two diagrams of the type of Fig. 1. Diagrammatically, doing contractions is represented by connecting diagrams in Eq. (79) while preserving the orientation of the lines. Since all operators in Eq. (74) have to be contracted, we have to perform all possible connections using Eq. (79). We can construct the diagrams (Fig. 2) which are redrawn in Fig. 3 in the usual form. The oriented lines are labelled with one index because the contractions (46) and (47) lead to Kronecker delta. Moreover, from Eqs. (46) and (47) we can specify which spin-orbitals are occupied and which are virtual.

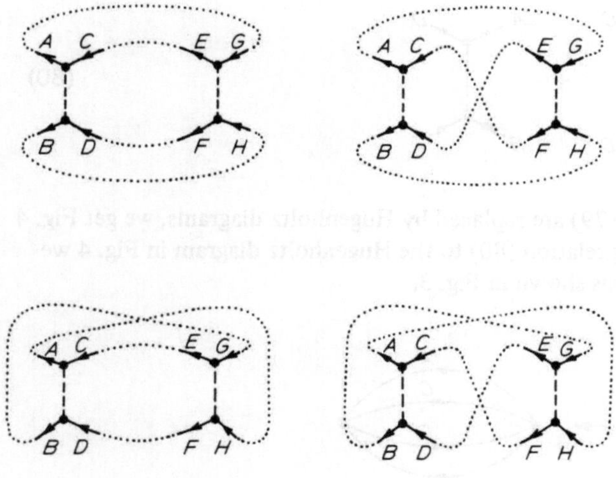


Fig. 2

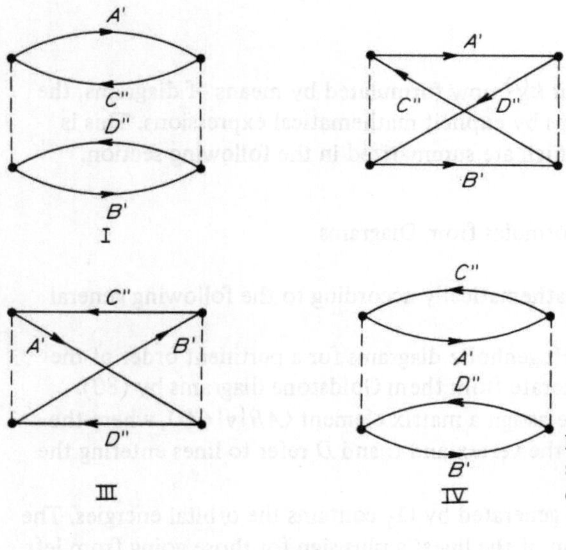


**Fig. 3. Goldstone diagrams for the second order contribution to the correlation energy in the ground state**

The diagrams in Fig. 3 are of the Goldstone type. The use of these diagrams is very advantageous because they allow us to obtain a very condense formula for the pertinent order of the perturbation expansion, and what is even more important, because the spin summation can be handled very simply. The disadvantage lies in the fact that for higher orders of perturbation theory the number of these diagrams increases rapidly and one must take care to ensure that all topologically distinct diagrams are constructed. Therefore, there is a need for a systematic procedure to guarantee that one has all the necessary diagrams. This can be achieved using the so-called Hugenholtz diagrams (or degenerate diagrams). The relation[37] between Hugenholtz and Goldstone diagrams is given by

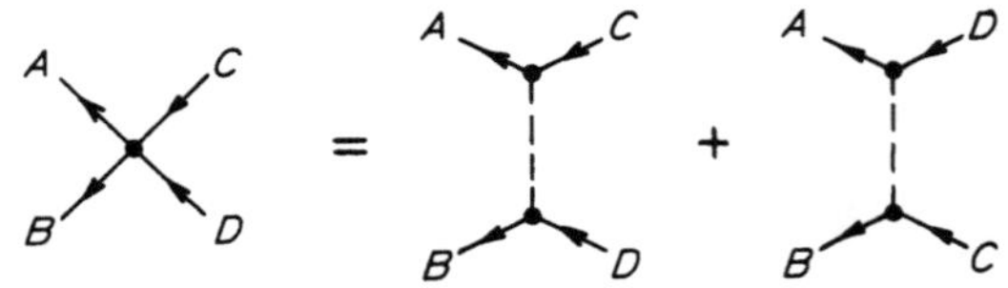


(80)

If Goldstone diagrams in Eq. (79) are replaced by Hugenholtz diagrams, we get Fig. 4 instead of Fig. 2. On applying relation (80) to the Hugenholtz diagram in Fig. 4 we obtain four Goldstone diagrams shown in Fig. 3.

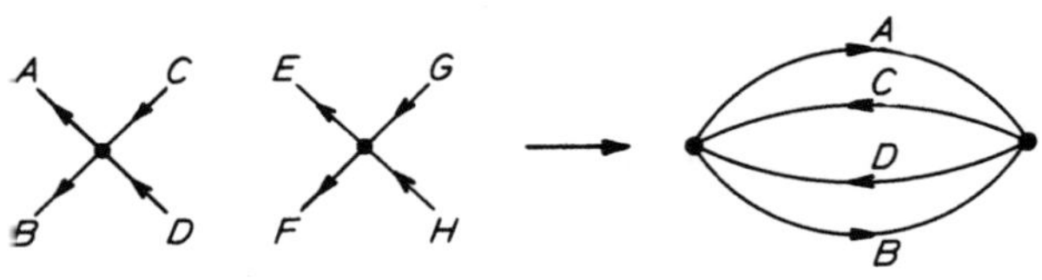


V

Fig. 4. Hugenholtz diagram for the second order contribution to the correlation energy in the ground state

Having the expression (74) for $k_0^{(2)}$ now formulated by means of diagrams, the next step is to replace the diagrams by explicit mathematical expressions. This is done according to certain rules which are summarized in the following section.

2. Rules for Obtaining Explicit Formulas from Diagrams

The diagrams can be expressed mathematically according to the following general rules:

1. Draw all possible distinct Hugenholtz diagrams for a pertinent order of the perturbation expansion[a)] and generate from them Goldstone diagrams by (80).

2. To each interaction line we assign a matrix element $\langle AB|v|CD\rangle$, where the indices $A, B$ refer to lines leaving the vertex and $C$ and $D$ refer to lines entering the vertex.

3. The denominator which is generated by $Q_0$ contains the orbital energies. The sign is determined by the direction of the lines: a plus sign for those going from left to right (hole lines) and a minus sign for those going from right to left (particle lines). A denominator factor is assigned to each pair of neighbouring interaction lines and the orbital energies are determined according to the oriented lines joining the two interaction lines.

4. The sign of the total expression corresponding to the pertinent Goldstone diagram is given by $(-1)^{l+h}$ where $l$ is the number of closed loops and $h$ is the number of hole lines.

5. Sum over all indices appearing in the pertinent Goldstone diagram.

a) At present, computer programs are available[38–40] for automatic generation of diagrams.

6. Sum the individual contributions of all Goldstone diagrams which are generated from the parent Hugenholtz diagram.

7. Multiply the whole expression (the sum of the Goldstone diagram contributions) with the topological factor $\frac{1}{2^n}$ where $n$ is the number of equivalent pairs of lines. An equivalent pair consists of two lines which both start at the same vertex of the Hugenholtz diagram and both end at the same vertex.

On applying the rules[b)] 1–7 to the problem of $k_0^{(2)}$ [Eq. (74), Diagrams I–IV in Fig. 3] we obtain

Diagram I:

$$+ \sum_{A'B'C''D''} \langle A'B'|v|C''D''\rangle \frac{1}{\epsilon_{A'} + \epsilon_{B'} - \epsilon_{C''} - \epsilon_{D''}} \langle C''D''|v|A'B'\rangle$$

Diagram II:

$$- \sum_{A'B'C''D''} \langle A'B'|v|C''D''\rangle \frac{1}{\epsilon_{A'} + \epsilon_{B'} - \epsilon_{C''} - \epsilon_{D''}} \langle D''C''|v|A'B'\rangle$$

Diagram III:

$$- \sum_{A'B'C''D''} \langle A'B'|v|C''D''\rangle \frac{1}{\epsilon_{A'} + \epsilon_{B'} - \epsilon_{C''} - \epsilon_{D''}} \langle C''D''|v|B'A'\rangle$$

Diagram IV:

$$+ \sum_{A'B'C''D''} \langle A'B'|v|C''D''\rangle \frac{1}{\epsilon_{A'} + \epsilon_{B'} - \epsilon_{C''} - \epsilon_{D''}} \langle C''D''|v|A'B'\rangle$$

We see that the Diagrams I and IV and Diagrams II and III respectively, give us the same contributions. The topological factor according to rule 7 is equal to $\frac{1}{4}$. Hence, the total expression becomes

$$k_0^{(2)} = \frac{1}{4} \sum_{A'B'C''D''} 2\langle A'B'|v|C''D''\rangle \frac{1}{\epsilon_{A'} + \epsilon_{B'} - \epsilon_{C''} - \epsilon_{D''}} \times \{\langle C''D''|v|A'B'\rangle - \langle D''C''|v|A'B'\rangle\} \tag{81}$$

b) The application of the rule 1 is trivial, in our case, because we have only one Hugenholtz diagram for $k_0^{(2)}$.

We recall that $|A'\rangle$, $|B'\rangle$, $|C''\rangle$, $|D''\rangle$ are spin-orbitals. To obtain the corresponding expression for orbitals we add the rule for the spin summation.

Rule 8: Since the spin must be conserved along the oriented line, it is necessary to introduce an additional numerical factor, $2^l$, with which we multiply each Goldstone diagram. As in rule 4, $l$ is the number of closed loops.

On applying this rule to Diagrams I to IV we obtain

Diagram I:

$$+4 \sum_{a'b'c''d''} \langle a'b'|v|c''d''\rangle \frac{1}{\epsilon_{a'} + \epsilon_{b'} - \epsilon_{c''} - \epsilon_{d''}} \langle c''d''|v|a'b'\rangle$$

Diagram II:

$$-2 \sum_{a'b'c''d''} \langle a'b'|v|c''d''\rangle \frac{1}{\epsilon_{a'} + \epsilon_{b'} - \epsilon_{c''} - \epsilon_{d''}} \langle d''c''|v|a'b'\rangle$$

Diagram III:

$$-2 \sum_{a'b'c''d''} \langle a'b'|v|c''d''\rangle \frac{1}{\epsilon_{a'} + \epsilon_{b'} - \epsilon_{c''} - \epsilon_{d''}} \langle c''d''|v|b'a'\rangle$$

Diagram IV:

$$+4 \sum_{a'b'c''d''} \langle a'b'|v|c''d''\rangle \frac{1}{\epsilon_{a'} + \epsilon_{b'} - \epsilon_{c''} - \epsilon_{d''}} \langle c''d''|v|a'b'\rangle$$

Then, adding up these contributions and multiplying them with the topological factor $\frac{1}{4}$ we arrive at the final expression

$$k_0^{(2)} = \sum_{a'b'c''d''} \langle a'b'|v|c''d''\rangle (2\langle c''d''|v|a'b'\rangle$$

$$- \langle d''c''|v|a'b'\rangle) \frac{1}{\epsilon_{a'} + \epsilon_{b'} - \epsilon_{c''} - \epsilon_{d''}} \tag{82}$$

Our derivation may appear strange due to the fact that we first introduced Goldstone diagrams into Eq. (79), then replaced these with Hugenholtz diagrams and finally again came back to Goldstone diagrams. Actually, the rules we are using here are the combined rules of Hugenholtz and Goldstone as suggested by Brandow[41]. The original Goldstone rules, which we are not presenting here, lead to complications with respect to ensuring that all topologically distinct diagrams are constructed, this being no easy task for higher orders of the perturbation expansions. Furthermore, we would have to determine the topological factor for each Goldstone diagram. It is understandable that the perturbation expansion for the quantity $k_i$ is not restricted to the ground state, *i.e.* $i \neq 0$, and in these cases the two disadvantages mentioned can lead to very laborious treatment. On the other hand, the spin summations and the determination of the correct sign factor can be correctly done only in Goldstone diagrams. A great advantage of Hugenholtz-type diagrams is that they are much fewer in number, each Hugenholtz diagram representing up to $2^n$ Goldstone diagrams, where $n$ is the number of vertices in the Hugenholtz diagram. The topological factor is given by a very simple rule (Rule 7), and we can generate all Goldstone diagrams by a systematic procedure [double projection scheme (80)]. For the reasons just noted we considered it reasonable to present these mixed Hugenholtz-Goldstone rules although the advantages of their use for the case of $k_0^{(2)}$ is not very evident. These mixed rules lead to construction of a larger number of Goldstone diagrams than would be obtained by strictly following Goldstone rules, however, this disadvantage is overweighed by the systematic nature of the generation of the diagrams as well as, by the convenience in the determination of their topological factors, especially for the cases $i \neq 0$.

For the sake of correctness, it is necessary to note that we disregarded such theoretically important concepts as the linked cluster (linked connected graph) theorem and the exclusion principle violating (EPV) diagrams. This is in accordance with our aim to maintain the practical nature of this review. The linked cluster theorem and EPV diagrams are of importance in the fourth and higher orders of the perturbation theory which, in our opinion, shall hardly be accessible to routine calculations in the foreseeable future. For detailed information on the linked cluster theorem and EPV diagrams see Refs.[9, 33, 34, 42].

## C. Third Order Contribution to the Correlation Energy

The expression (73) can be schematically written [similar to Eq. (79)] in the form

$k_0^{(3)} = \langle\Phi_0|$  $Q_0$  $Q_0$  $|\Phi_0\rangle$ (83)

Here we can construct three different Hugenholtz-type diagrams (Fig. 5)

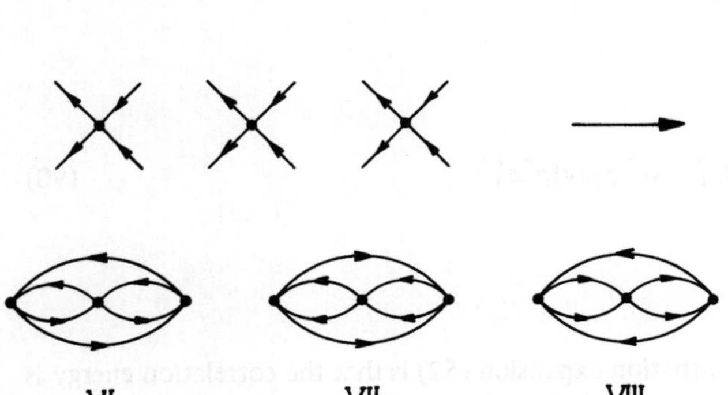

VI VII VIII

Fig. 5. Hugenholtz diagrams for the third order contribution to the correlation energy in the ground state

Following the rules 1–8 one can arrive at the explicit expressions for the Diagrams VI–VIII and a very compact formula[9)] for $k_0^{(3)}$. We limit ourselves here to the final formula (for details of its derivation see Ref.[9)])

$$k_0^{(3)} = \sum_{\substack{a',b',c'\\a'',b'',c''}} R_{\mathrm{ph}}[(2A_0 - A_1)(2B_0 - B_1)(2C_0 - C_1) - 3A_1B_1C_1]$$

$$+ \sum_{\substack{a_1',a_2'\\b_1'',b_2'',c_1'',c_2''}} R_{\mathrm{pp}} A_0' B_0' (2C_0' - C_1') + \sum_{\substack{b_1',b_2',c_1',c_2'\\a_1'',a_2''}} R_{\mathrm{hh}} A_0'' B_0'' (2C_0'' - C_1'') \qquad (84)$$

where $R_{\mathrm{ph}}, R_{\mathrm{pp}}$ and $R_{\mathrm{hh}}$ are the denominator factors (subscripts p and h stand for particles and holes, respectively)

$$R_{\mathrm{ph}} = [(\epsilon_{a'} + \epsilon_{c'} - \epsilon_{a''} - \epsilon_{c''})(\epsilon_{a'} + \epsilon_{b'} - \epsilon_{a''} - \epsilon_{b''})]^{-1} \qquad (85)$$

$$R_{\mathrm{pp}} = [(\epsilon_{a_1'} + \epsilon_{a_2'} - \epsilon_{c_1''} - \epsilon_{c_2''})(\epsilon_{a_1'} + \epsilon_{a_2'} - \epsilon_{b_1''} - \epsilon_{b_2''})]^{-1} \qquad (86)$$

$$R_{\mathrm{hh}} = [(\epsilon_{c_1'} + \epsilon_{c_2'} - \epsilon_{a_1''} - \epsilon_{a_2''})(\epsilon_{b_1'} + \epsilon_{b_2'} - \epsilon_{a_1''} - \epsilon_{a_2''})]^{-1} \qquad (87)$$

The $A, B, C$ terms are matrix elements of the v-operator (electronic repulsion integrals) over orbitals. The particle-hole terms are

$$\begin{aligned} A_0 &= \langle a''b''|v|a'b'\rangle & A_1 &= \langle a''b''|v|b'a'\rangle \\ B_0 &= \langle b'c''|v|b''c'\rangle & B_1 &= \langle b'c''|v|c'b''\rangle \\ C_0 &= \langle a'c'|v|a''c''\rangle & C_1 &= \langle a'c'|v|c''a''\rangle \end{aligned} \qquad (88)$$

the particle-particle terms are

$$\begin{aligned} A_0' &= \langle b_1''b_2''|v|a_1'a_2'\rangle \\ B_0' &= \langle c_1''c_2''|v|b_1''b_2''\rangle \\ C_0' &= \langle a_1'a_2'|v|c_1''c_2''\rangle \qquad C_1' = \langle a_1'a_2'|v|c_2''c_1''\rangle \end{aligned} \qquad (89)$$

and the hole-hole terms are

$$\begin{aligned} A_0'' &= \langle a_1''a_2''|v|b_1'b_2'\rangle \\ B_0'' &= \langle b_1'b_2'|v|c_1'c_2'\rangle \\ C_0'' &= \langle c_1'c_2'|v|a_1''a_2''\rangle \qquad C_1'' = \langle c_1'c_2'|v|a_2''a_1''\rangle \end{aligned} \qquad (90)$$

## D. Infinite Summations

A typical feature of the perturbation expansion (52) is that the correlation energy is expressed by way of an infinite series. It is understandable that for actual calculations the expansion (52) must be truncated. It is one of the outstanding advantages of the many-body theory that it allows one to sum certain types of diagram contributions up to the infinite order. In this section we are going to present a certain type of infinite summation. The idea of this technique, which is called the ladder technique, comes from nuclear physics[43] and was used for the first time in quantum chemistry by Kelly[44, 45]. We shall demonstrate this technique with of aid of Diagram I (of Fig. 3). In this diagram we have two interaction lines. The idea of ladders is to try to extend the number of interaction lines to infinity. For example, we might try to write an explicit expression for the Diagrams IX–XI in Fig. 6. Note that we pre-

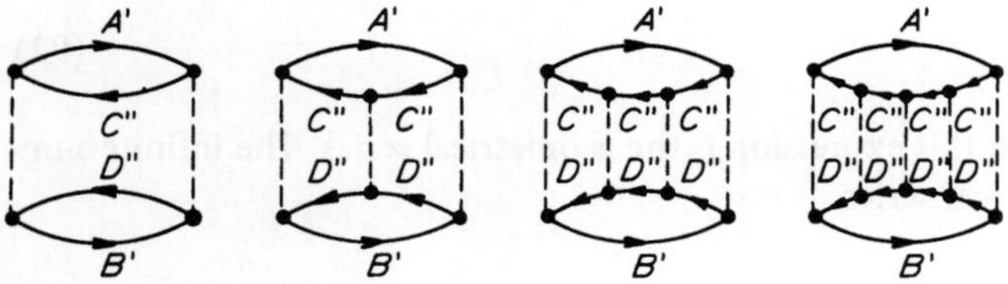


I IX X XI

Fig. 6. Evolution of the diagonal ladder diagrams

serve the same index between added interaction lines. The diagrams of this type are called the diagonal ladders[c]. We see that these diagrams correspond to different orders of the perturbation expansion. We might try to calculate the sum of the contributions of these diagrams to the infinite order of the perturbation expansion. We can achieve this by defining the new interaction line as

$$\wr = \sum_{i=0}^{\infty} [\,\vdots\,]^i \tag{91}$$

Our aim is therefore to write the explicit expression for the following diagram

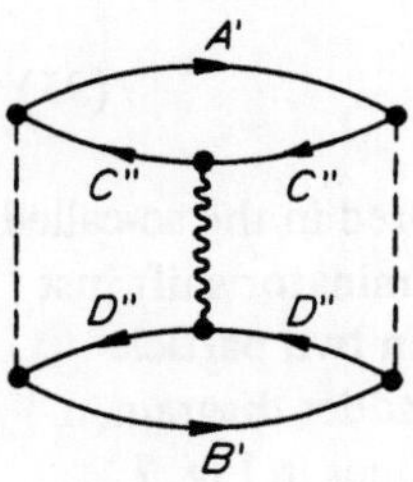


XII Fig. 7

c) In the time-dependent theory the diagrams of the type in Diagram XI are drawn in such a way that they indeed resemble a ladder.

This can be done following the rules of Section IV.B.2, but remembering that we have a new interaction line defined by (91). Thus, the Diagram XII in Fig. 7 gives us

$$\sum_{A'B'C''D''} \langle A'B'|v|C''D''\rangle \frac{1}{\epsilon_{A'} + \epsilon_{B'} - \epsilon_{C''} - \epsilon_{D''}}$$
$$\times \left\{ \sum_{i=0}^{\infty} \left[ \langle C''D''|v|C''D''\rangle \frac{1}{\epsilon_{A'} + \epsilon_{B'} - \epsilon_{C''} - \epsilon_{D''}} \right]^i \right\}$$
$$\times \langle C''D''|v|A'B'\rangle \tag{92}$$

The infinite summation in this expression is the geometrical series. The infinite summation ($S_n$) of a geometrical series

$$a + aq + aq^2 + aq^3 + \ldots$$

is given by

$$S_n = \frac{a}{1-q} \tag{93}$$

In our case this means that

$$S_n = \frac{1}{1 - \dfrac{\langle C''D''|v|C''D''\rangle}{\epsilon_{A'} + \epsilon_{B'} - \epsilon_{C''} - \epsilon_{D''}}} = \frac{\epsilon_{A'} + \epsilon_{B'} - \epsilon_{C''} - \epsilon_{D''}}{\epsilon_{A'} + \epsilon_{B'} - \epsilon_{C''} - \epsilon_{D''} - \langle C''D''|v|C''D''\rangle} \tag{94}$$

Substituting (94) in (92) gives us the expression:

$$\sum_{A'B'C''D''} \langle A'B'|v|C''D''\rangle \frac{1}{\epsilon_{A'} + \epsilon_{B'} - \epsilon_{C''} - \epsilon_{D''} - \langle C''D''|v|C''D''\rangle}$$
$$\times \langle C''D''|v|A'B'\rangle \tag{95}$$

We see that the whole effect of the infinite summation (91) appeared in the so-called denominator shift [compare expressions (81) and (95)]. The denominator shift just presented, corresponds to putting the interaction line (91) between two particle lines. For this reason we talk about the diagonal particle-particle ladder diagram. Of course, we also have other possibilities of joining the oriented lines in Fig. 7. These provide ways we can arrive at the hole-particle and hole-hole denominator shifts (DS). By adding up all possible contributions noted, we arrive at the final expression

$$\sum_{A'B'C''D''} \langle A'B'|v|C''D''\rangle \frac{1}{\epsilon_{A'} + \epsilon_{B'} - \epsilon_{C''} - \epsilon_{D''} + DS}$$
$$\times \langle C''D''|v|A'B'\rangle \tag{96}$$

where

$$DS = -\langle C''D''|v|C''D''\rangle_A - \langle A'B'|v|A'B'\rangle_A + \langle A'C''|v|A'C''\rangle_A + \langle A'D''|v|A'D''\rangle_A + \langle B'C''|v|B'C''\rangle_A + \langle B'D''|v|B'D''\rangle_A \tag{97}$$

The antisymmetrized matrix elements in Eq. (97) are defined as follows:

$$\langle AB|v|CD\rangle_A = \langle AB|v|CD\rangle - \langle AB|v|DC\rangle \tag{98}$$

As was shown by Claverie *et al.*[46] this denominator shift leads to the same expression for $k_0$ as does the Epstein-Nesbet partitioning of the Hamiltonian.

## V. Cluster Expansion of the Wave Function

### A. Brief Description

In this section we shall discuss an approach which is neither variational nor perturbational. This approach also has its origin in nuclear physics and was introduced to quantum chemistry by Sinanoğlu[47]. It is based on a cluster expansion of the wave function. A systematic method for the calculation of cluster expansion components of the exact wave function was developed by Čížek[48]. The characteristic feature of this approach is the expansion of the wave function as a linear combination of Slater determinants. Formally, this expansion is similar to the ordinary CI expansion. The cluster expansion, however, gives us not only the physical insight of the correlation energy but it also shows the connections between the variational approaches (CI) and the perturbational approaches (*e.g.* MB-RSPT).

We shall express the exact wave function in the form

$$|\Psi\rangle = |\Phi_0\rangle + |\eta\rangle \tag{99}$$

where $|\Phi_0\rangle$ is the Slater determinant for the closed shell ground state, and $|\eta\rangle$ is the correlation function, which describes the correlation of two, three *etc.* electrons. We can choose the correlation function of a $N$-electron system in the form

$$|\eta\rangle = \sum_i |U^{(i)}\rangle + \sum_{i<j} |U^{(ij)}\rangle + \sum_{i<j<k} |U^{(ijk)}\rangle + \sum_{i<j<k<l} |U^{(ijkl)}\rangle + \ldots \tag{100}$$

where the indices $i, j, k, l \ldots$ are indices of the electrons 1 to $N$. The functions $|U^{(i)}\rangle$ are called one-electron clusters, $|U^{(ij)}\rangle$-two-electron clusters, $|U^{(ijk)}\rangle$-three-electron clusters, *etc.* Let us examine in more detail the term $|U^{(ijkl)}\rangle$ which describes the correlation of four electrons. We can distinguish the simultaneous correlation of four electrons (so called linked cluster), for example, from the interaction of two pairs of electrons (so called unlinked cluster). Simultaneous four electron correlation occurs only in situations when all four electrons are close together. Since such "collisions" are rare in molecules we can expect that the effect of linked four electron clusters in expansion (100) will not be important[47]. On the other hand, the effect of unlinked clusters may be important since these correspond to a collision of two electrons $i$ and $j$ and the independent and simultaneous collision of another two electrons $k$ and $l$. Obviously, collisions of another type are also conceivable and any cluster can therefore be described as the sum of linked and unlinked clusters

$$
\begin{aligned}
|U^{(i)}\rangle &= \mathrm{t}^{(i)}|\Phi_0\rangle \\
|U^{(ij)}\rangle &= \mathrm{t}^{(ij)}|\Phi_0\rangle + \mathrm{t}^{(i)}\mathrm{t}^{(j)}|\Phi_0\rangle \\
|U^{(ijk)}\rangle &= \mathrm{t}^{(ijk)}|\Phi_0\rangle + \mathrm{t}^{(k)}\mathrm{t}^{(ij)}|\Phi_0\rangle \\
&\quad + \mathrm{t}^{(j)}\mathrm{t}^{(ik)}|\Phi_0\rangle + \mathrm{t}^{(i)}\mathrm{t}^{(jk)}|\Phi_0\rangle \\
&\quad + \mathrm{t}^{(i)}\mathrm{t}^{(j)}\mathrm{t}^{(k)}|\Phi_0\rangle
\end{aligned}
\tag{101}
$$

*etc.*

In Eqs. (101) the t operators are generating the clusters $|U\rangle$ from the Slater determinant $|\Phi_0\rangle$. Expansion (99) by means of t operators as indicated in Eqs. (101) can be expressed in the compact form[49]

$$|\Psi\rangle = e^{\mathrm{T}}|\Phi_0\rangle \tag{102}$$

Expanding $e^{\mathrm{T}}$ we get

$$|\Psi\rangle = \left\{1 + \mathrm{T} + \frac{\mathrm{T}^2}{2!} + \frac{\mathrm{T}^3}{3!} + \ldots\right\}|\Phi_0\rangle \tag{103}$$

T is generally an $N$-electron operator

$$\mathrm{T} = \sum_n \mathrm{T}_n; \quad n = 1, 2, \ldots, N \tag{104}$$

where

$$
\begin{aligned}
\mathrm{T}_1 &= \sum_i \mathrm{t}^{(i)} \\
\mathrm{T}_2 &= \sum_{i<j} \mathrm{t}^{(ij)}
\end{aligned}
\tag{105}
$$

*etc.*

Let us compare our cluster expansion with the well known CI expansion of the wave function

$$|\Psi\rangle = C_0|\Phi_0\rangle + \sum_{AR}\sum C_A^R|\Phi_A^R\rangle + \sum_{A<B}\sum_{R<S} C_{AB}^{RS}|\Phi_{AB}^{RS}\rangle + \ldots \tag{106}$$

Indices $A, B$ belong to occupied spin-orbitals and $R, S$ to virtual spin-orbitals. The determinant $|\Phi_{AB}^{RS}\rangle$ is a doubly excited configuration. Let us normalize the wave function (106), by setting $C_0 = 1$, so as to obtain a form compatible with the wave function (99):

$$|\Psi\rangle = |\Phi_0\rangle + \mathrm{C}_1 |\Phi_0\rangle + \mathrm{C}_2 |\Phi_0\rangle + \ldots \tag{107}$$

Here the operators $\mathrm{C}_n$ generate all $n$-excited configurations. Comparing Eqs. (107) and (103) we can write[50, 51]

$$\mathrm{C}_n = \mathrm{T}_n + \mathrm{Q}_n \tag{108}$$

where $\mathrm{Q}_n$ contains the products of $\mathrm{T}_m$ operators for $m < n$;

$$\mathrm{Q}_1 = 0 \tag{109}$$

$$\mathrm{Q}_2 = \frac{1}{2!}\mathrm{T}_1^2 \tag{110}$$

$$\mathrm{Q}_3 = \mathrm{T}_1\mathrm{T}_2 + \frac{1}{3!}\mathrm{T}_1^3 \tag{111}$$

$$\mathrm{Q}_4 = \mathrm{T}_1\mathrm{T}_3 + \frac{1}{2}\mathrm{T}_1^2\mathrm{T}_2 + \frac{1}{4!}\mathrm{T}_1^4 + \frac{1}{2!}\mathrm{T}_2^2 \tag{112}$$

The Q operators generate the unlinked clusters.

In the second quantization formalism the components of the T operator (104) may be expressed as follows

$$\mathrm{T}_1 = \sum_{AR}\sum d_A^R \mathrm{X}_R^+ \mathrm{X}_A \tag{113}$$

$$\mathrm{T}_2 = \sum_{A<B}\sum_{R<S} d_{AB}^{RS} \mathrm{X}_R^+ \mathrm{X}_S^+ \mathrm{X}_A \mathrm{X}_B \tag{114}$$

*etc.* where the $d$'s are coefficients in the cluster expansion (102). This expression permits us a very elegant comparison[8] of the wave functions in the cluster with those of the CI expansions. Comparing Eqs. (103) and (107) and assuming relations (104), (113) and (114) we obtain

$$C_A^R = d_A^R \tag{115}$$

$$C_{AB}^{RS} = d_{AB}^{RS} + d_A^R d_B^S - d_A^S d_B^R \tag{116}$$

$$C_{ABC}^{RST} = d_{ABC}^{RST} + d_A^R d_{BC}^{ST} - d_B^R d_{AC}^{ST} + 7 \text{ permuted terms} + d_A^R d_B^S d_C^T + 5 \text{ permuted terms} \tag{117}$$

In the introduction to this section we made mention of the clusters $\mathrm{t}^{(ijkl)}|\Phi_0\rangle$ and $\mathrm{t}^{(ij)}\mathrm{t}^{(kl)}|\Phi_0\rangle$. For further consideration it is very useful to also make some remarks concerning other clusters. The contribution from $\mathrm{t}^{(i)}|\Phi_0\rangle$ will be small when

Hartree-Fock orbitals are used. The contributions from these clusters are exactly equal to zero when Brueckner orbitals are used[50]. It is now also understandable that all products containing $t^{(i)}$ such as $t^{(i)}t^{(j)}$, $t^{(i)}t^{(jk)}$ *etc.* will be small. Generally, the effect of linked tri-excited clusters as well as that of linked tetra-excited clusters is small. (The calculations demonstrating these effects will be discussed in Section VI.G.) It is possible to state that from three-electron clusters upwards the contribution to the correlation energy coming from linked clusters decreases rapidly. This is also understandable due to the fact that these contributions first appear in terms of higher orders in the perturbation expansion, which is demonstrated in Table 1. From the above discussion it is possible to conclude that the most important clusters are

$T_2 | \Phi_0 \rangle$ and $\frac{1}{2} T_2^2 | \Phi_0 \rangle$.

Table 1. Lowest order of perturbation theory (LOPT) in which various linked and unlinked clusters first appear[51]

| Linked clusters | LOPT | Unlinked clusters | LOPT |
|---|---|---|---|
| $T_1$ | 2 | $T_1 T_2$ | 3 |
| | | $\frac{1}{2} T_1^2$ | 4 |
| $T_2$ | 1 | $\frac{1}{2} T_2^2$ | 2 |
| $T_3$ | 2 | $T_1 T_3$ | 4 |

### B. Correlation Energy Through Cluster Expansion

Our goal is to solve again the equation

$$H | \Psi \rangle = E | \Psi \rangle \tag{118}$$

where $| \Psi \rangle$ is the exact wave function and $E$ is the exact energy of a $N$-electron system. Let us substitute the expansion (102) into Eq. (118) which gives us

$$H \, e^{T} | \Phi_0 \rangle = E \, e^{T} | \Phi_0 \rangle \tag{119}$$

From the discussion in Section V.A. it appears that it is reasonable to assume $T \approx T_2$. Equation (119) will be then

$$H \, e^{T_2} | \Phi_0 \rangle = E \, e^{T_2} | \Phi_0 \rangle \tag{120}$$

or

$$H\left\{1 + T_2 + \frac{T_2^2}{2!} + \ldots\right\}|\Phi_0\rangle = E\left\{1 + T_2 + \frac{T_2^2}{2!} + \ldots\right\}|\Phi_0\rangle \tag{121}$$

Let us substract the term $\langle\Phi_0 | H | \Phi_0\rangle$ from both sides of Eq. (121). This gives us

$$\{H - \langle\Phi_0|H|\Phi_0\rangle\}\left\{1 + T_2 + \frac{T_2^2}{2} + \ldots\right\}|\Phi_0\rangle =$$

$$= \{E - \langle\Phi_0|H|\Phi_0\rangle\}\left\{1 + T_2 + \frac{T_2^2}{2} + \ldots\right\}|\Phi_0\rangle \tag{122}$$

The term $\{E - \langle\Phi_0|H|\Phi_0\rangle\}$ is the correlation energy of the closed shell ground state. As was shown in Section III the two-particle operator $T_2$ can be expressed in the second quantization formalism as

$$T_2 = \frac{1}{2}\sum_{EFGH} \langle EF|t_2|GH\rangle N[X_E^+ X_F^+ X_H X_G] \tag{123}$$

Operator $t_2$ has the same meaning as the $t^{(ij)}$ operators in Eq. (105). The term on the right hand side of Eq. (123) closely resembles the expression (67). In analogy to (67), we can also represent this term diagrammatically in the following way:

$$\longrightarrow \langle EF|t_2|GH\rangle N[X_E^+ X_F^+ X_H X_G] \tag{124}$$

The quantity $\{H - \langle\Phi_0|H|\Phi_0\rangle\}$, we already know, is our Hamiltonian in the normal product form (61). The expansion

$$e^T = 1 + T_2 + \frac{T_2^2}{2} + \ldots$$

can be represented schematically by diagrams as

$e^T = 1 +$  $+$  $+ \cdots$ (125)

Therefore Eq. (122) can also be schematically represented by the diagrams

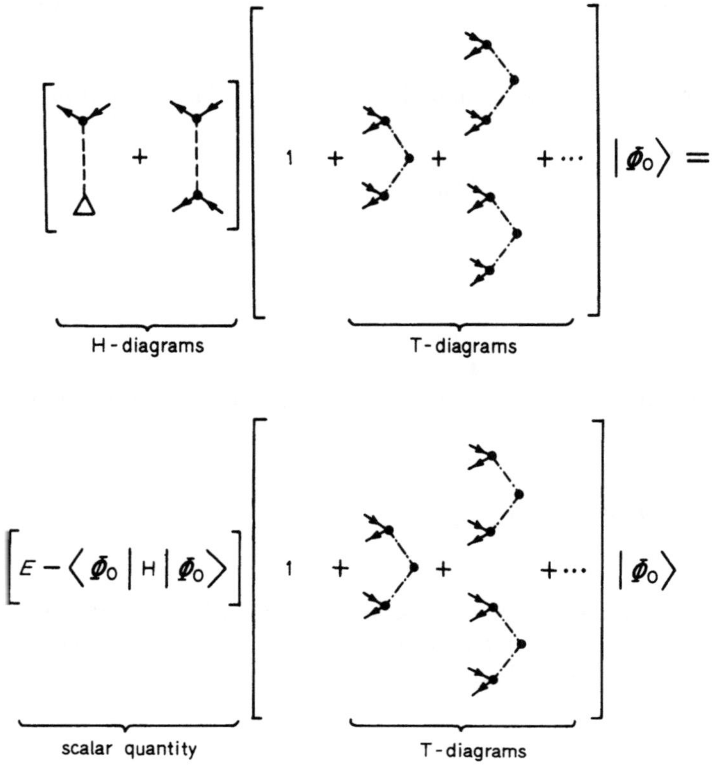


(126)

The meaning of all these diagrams with the exception of the first diagram on the left hand side of Eq. (126) is already known to us. This diagram has the following meaning

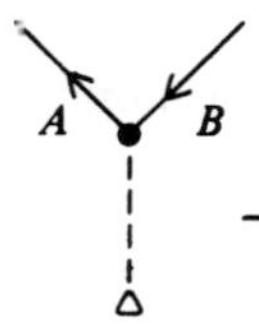


$$\longrightarrow \sum_{AB} \langle A|f|B\rangle N\left[X_A^+ X_B\right] \tag{127}$$

and originates from the normal product form of the Hamiltonian (61).

According to the generalized Wick's theorem, we can form new diagrams ($R$-diagrams in notation of Ref.[48]) on the left hand side of Eq. (126) that connect the $H$- and $T$-diagrams. We can distinguish two types of $R$-diagrams:

(i) those where all lines are connected (we say these diagrams have no open path),

(ii) those where not all lines are connected (which we shall call diagrams with an open path).

It should be also noted that the $R$-diagrams can be either connected or disconnected, the term "disconnected" is assigned to diagrams which contain two or more parts (see Fig. 8). Upon inspection of Eq. (126) it may be seen that by connecting *H*- and *T*-diagrams we get both connected and disconnected *R*-diagrams. Furthermore, it may be noticed that the rests of the resulting disconnected diagrams, form the same series on the left hand side of Eq. (126), as do the *T*-diagrams on the right hand side, their sum therefore being equal to $e^{T_2}$. By rests of disconnected diagrams we mean that which is schematically shown in Fig. 9, below the dashed line. There-

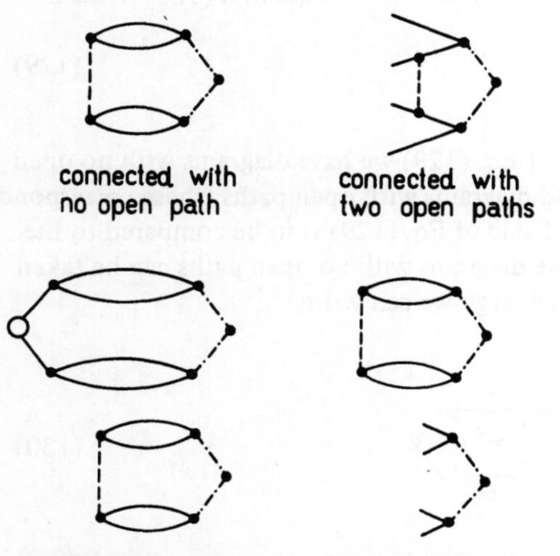


Fig. 8. Types of *R*-diagrams

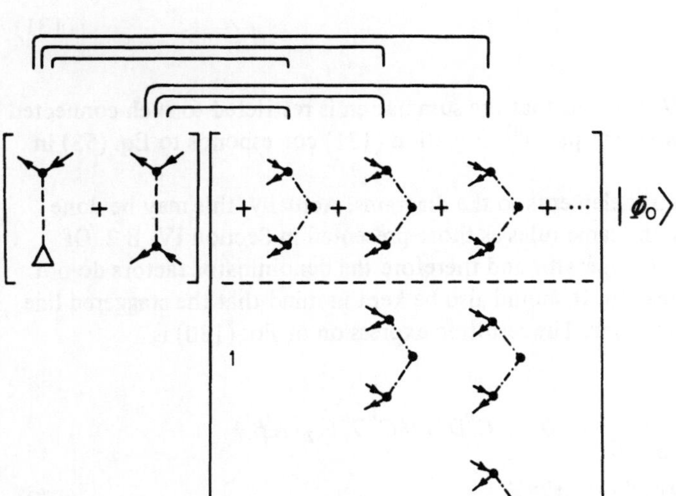


Fig. 9. Schematic representation of connections between the *H* and *T* diagrams on the left hand side of Eq. (121). The terms below the dashed line represent expanded $e^{T_2}$

fore we can extract $e^{T_2}$ from both sides of Eq. (126) and rewrite Eq. (126) in the form

$$e^{T_2}\{\sum_r [R(r)]_c - (E - \langle\Phi_0|H|\Phi_0\rangle)\} = 0 \tag{128}$$

where the subscript c means that we sum only connected diagrams. We note that Eq. (128) is completely equivalent to Eq. (51) in Ref.[48]. Equation (128) holds if

$$\sum_r [R(r)]_c = E - \langle\Phi_0|H|\Phi_0\rangle \tag{129}$$

We point out that on the left hand side of Eq. (129) we have diagrams with no open path (these give us a scalar quantity) and diagrams with open paths (these correspond to operators). Therefore, if the left hand side of Eq. (129) is to be compared to the scalar quantity $E - \langle\Phi_0|H|\Phi_0\rangle$, only the diagrams with no open paths can be taken into account. Hence, for the correlation energy we can write

$E - \langle\Phi_0|H|\Phi_0\rangle =$ 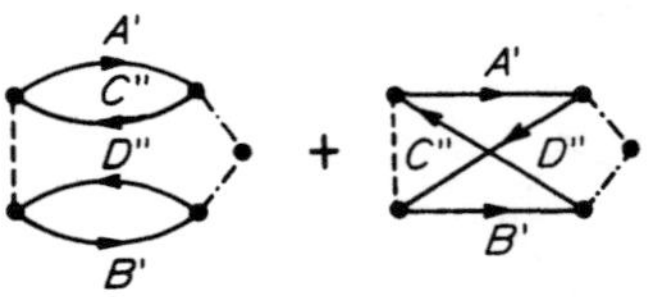 (130)


This equation is equivalent to Eq. (52) in Ref.[48]. Since the diagrams with open paths have no counterpart on the right hand side of Eq. (129) it holds that

$$\sum_r [R(r)]_c^{O.P.} = 0 \tag{131}$$

where the superscript *O.P.* means that the summation is restricted to such connected diagrams which have two open paths[d]. Equation (131) corresponds to Eq. (53) in Ref.[48].

Next, we assign matrix elements to the diagrams. Actually, this may be done according to essentially the same rules as those presented in Section IV. B.2. Of course, here we have no $Q_i$ operator and therefore the denominator factors do not appear in the final expression. It should also be kept in mind that the staggered line corresponds to the $t_2$ operator. The resulting expression of Eq. (130) is

$$E - \langle\Phi_0|H|\Phi_0\rangle = \frac{1}{2}\sum_{A'B'C''D''}\langle A'B'|v|C''D''\rangle\{\langle C''D''|t_2|A'B'\rangle - \langle D''C''|t_2|A'B'\rangle\} \tag{132}$$

[d] This is due to the fact that T is set equal to $T_2$. Note that the diagrams with one open path cannot be constructed without a violation of the generalized Wick's theorem.

or in the orbital form

$$E - \langle \Phi_0 | \mathsf{H} | \Phi_0 \rangle = \sum_{a'b'c''d''} \langle a'b' | \mathsf{v} | c''d'' \rangle \{ 2 \langle c''d'' | \mathsf{t}_2 | a'b' \rangle - \langle d''c'' | \mathsf{t}_2 | a'b' \rangle \} \tag{133}$$

We see that the correlation energy (133) is given by the matrix elements of v and $t_2$ operators. The former are the electronic repulsion integrals which can be calculated from the known Hartree-Fock orbitals. The matrix elements of $t_2$ operators are unknown but have to the determined in order to calculate the correlation energy (133). It is just Eq. (131) which leads to a system of nonlinear equations whose solutions provides us with the necessary $t_2$-matrix elements.

We shall now briefly show how to construct the corresponding diagrams with two open paths. Of course, we have to remember that in addition to the v-interaction line and the $t_2$-interaction line, we still have another new interaction line corresponding to the f-operator [see Eq. (127)]. Any diagram containing the f-operator gives a vanishing expression unless the entering and leaving lines in the f-part of the diagram have the same direction and a common index, this due to the behaviour of Hartree-Fock orbitals

$$\langle a'' | \mathsf{f} | b'' \rangle = \delta_{a''b''} \epsilon_{a''} \tag{134}$$

$$\langle a' | \mathsf{f} | b' \rangle = \delta_{a'b'} \epsilon_{a'} \tag{135}$$

For the construction of diagrams with two open paths we have the possibilities shown in Figs. 10–12. The diagrams containing f and $t_2^2$ operators are not presented because the contributions of these diagrams are equal to zero when using the Hartree-Fock orbitals.

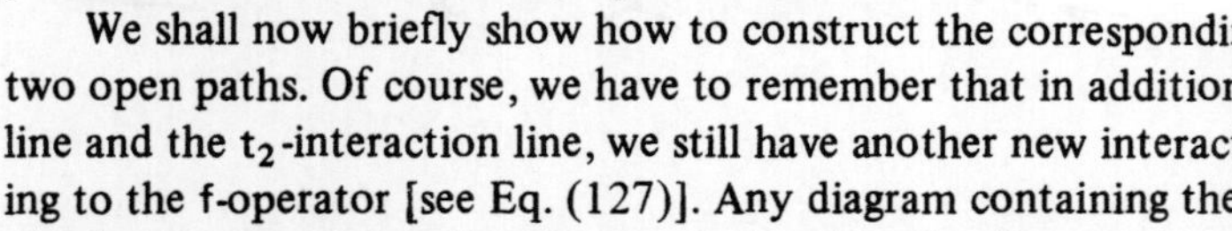

Fig. 10. Diagrams with two open paths which contain f and $t_2$ operators

The assignment of explicit expressions to diagrams in Figs. 10–12 is made in the same manner as for diagrams already mentioned. For the sake of illustration we present the assignment of the first diagram in Fig. 10.

$$\longrightarrow \sum_{ABCDE} \langle C'' | \mathsf{f} | B'' \rangle \langle B''E'' | \mathsf{t}_2 | A'D' \rangle \tag{136}$$

Correctly speaking, expression (136) should also contain the normal product $N[X_A^+ X_B^+ X_E X_D]$ as given by relationship (124). This term is irrelevant, however,

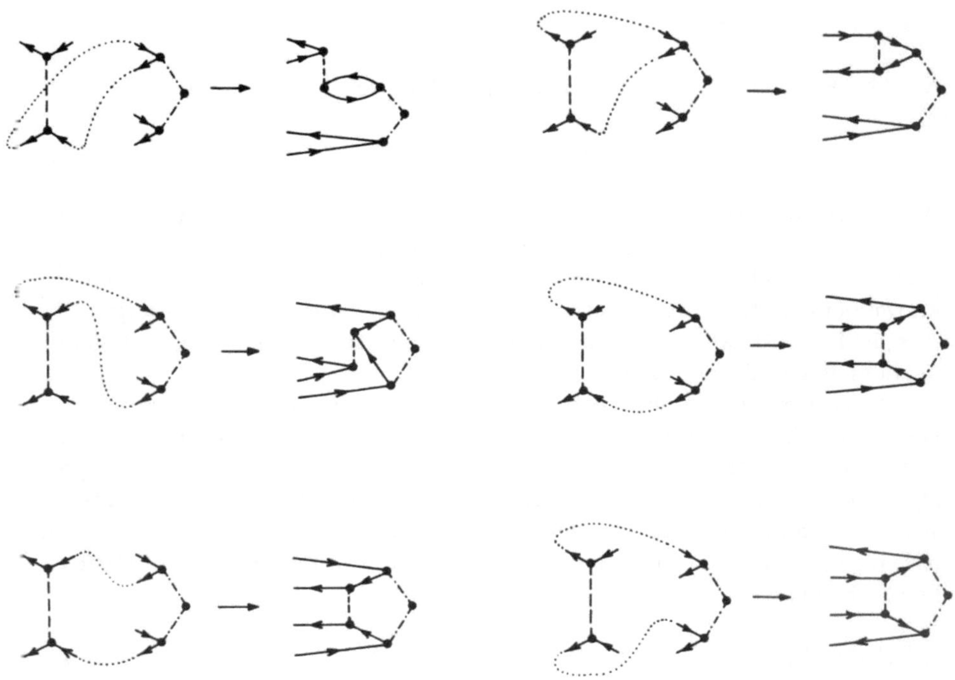

Fig. 11. Diagrams with two open paths which contain v and $t_2$ operators

for the determination of the final form of Eq. (131), and is therefore here omitted for simplicity. The final equations determining $t_2$ operators are obtained from the summation of individual contributions from all pertinent diagrams in Figs. 10–12. The equations have the following form[48)]

$$\lambda(d''_1, d'_1; d''_2, d'_2) + \lambda(d''_2, d'_2; d''_1, d'_1) = 0 \tag{137}$$

where

$$\begin{aligned}
\lambda(d''_1, d'_1; d''_2, d'_2) &= \frac{1}{2} \langle d''_1 d''_2 | \mathrm{v} | d'_1 d'_2 \rangle + \sum_{d''_3} \langle d''_1 | \mathrm{f} | d''_3 \rangle \langle d''_3 d''_2 | \mathrm{t}_2 | d'_1 d'_2 \rangle \\
&- \sum_{d'_3} \langle d'_1 | \mathrm{f} | d'_3 \rangle \langle d''_1 d''_2 | \mathrm{t}_2 | d'_3 d'_2 \rangle + \sum_{d'_3, d''_3} [(2 \langle d''_1 d'_3 | \mathrm{v} | d'_1 d''_3 \rangle \\
&- \langle d''_1 d'_3 | \mathrm{v} | d''_3 d'_1 \rangle) \langle d''_3 d''_2 | \mathrm{t}_2 | d'_3 d'_2 \rangle - \langle d''_1 d'_3 | \mathrm{v} | d'_1 d''_3 \rangle \langle d''_2 d''_3 | \mathrm{t}_2 | d'_3 d'_2 \rangle \\
&- \langle d''_2 d'_3 | \mathrm{v} | d''_3 d'_1 \rangle \langle d''_1 d''_3 | \mathrm{t}_2 | d'_3 d'_2 \rangle] \\
&+ \frac{1}{2} \sum_{d''_3, d''_4} \langle d''_1 d''_2 | \mathrm{v} | d''_3 d''_4 \rangle \langle d''_3 d''_4 | \mathrm{t}_2 | d'_1 d'_2 \rangle + \frac{1}{2} \sum_{d'_3, d'_4} \langle d'_1 d'_2 | \mathrm{v} | d'_3 d'_4 \rangle \langle d''_1 d''_2 | \mathrm{t}_2 | d'_3 d'_4 \rangle \\
&+ \sum_{d'_3, d'_4; d''_3, d''_4} \{[2 \langle d'_3 d'_4 | \mathrm{v} | d''_3 d''_4 \rangle - \langle d'_3 d'_4 | \mathrm{v} | d''_4 d''_3 \rangle] [\langle d''_1 d''_3 | \mathrm{t}_2 | d'_1 d'_3 \rangle \langle d''_2 d''_4 | \mathrm{t}_2 | d'_2 d'_4 \rangle
\end{aligned}$$

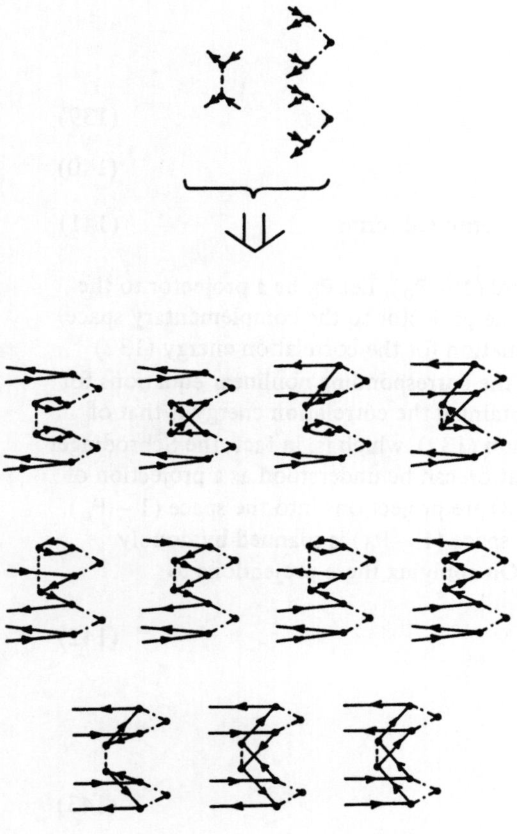

Fig. 12. Diagrams with two open paths which contain v and $t_2^2$ operators

$$- \langle d''_1 d''_3 | t_2 | d'_3 d'_1 \rangle \langle d''_2 d''_4 | t_2 | d'_2 d'_4 \rangle - \langle d''_3 d''_4 | t_2 | d'_3 d'_1 \rangle \langle d''_1 d''_2 | t_2 | d'_4 d'_2 \rangle$$

$$- \langle d''_1 d''_3 | t_2 | d'_4 d'_3 \rangle \langle d''_2 d''_4 | t_2 | d'_2 d'_1 \rangle] + \frac{1}{2} \langle d'_3 d'_4 | v | d''_3 d''_4 \rangle [\langle d''_1 d''_3 | t_2 | d'_3 d'_1 \rangle \langle d''_2 d''_4 | t_2 | d'_4 d'_2 \rangle$$

$$+ \langle d''_2 d''_3 | t_2 | d'_4 d'_1 \rangle \langle d''_1 d''_4 | t_2 | d'_3 d'_2 \rangle + \langle d''_1 d''_2 | t_2 | d'_3 d'_4 \rangle \langle d''_3 d''_4 | t_2 | d'_1 d'_2 \rangle]\}. \qquad (138)$$

The method just described is known as the CPMET method. This approach, based on the cluster expansion, is quite general and permits the inclusion of arbitrary clusters into the $e^T$ expansion. This will be discussed in Section VI.G.

To further a better understanding of the method, we now demonstrate a simple derivation of CPMET equations in a more traditional manner according to Kutzelnigg[8]. It should be emphasized however, that this approach is applicable only in the simple case of $e^T \approx e^{T_2}$. On applying the restriction $T \approx T_2$, the comparison of cluster and CI expansions represented by Eqs. (115)–(117) reduces to

$$C_{AB}^{RS} = d_{AB}^{RS} \tag{139}$$

$$C_{ABC}^{RST} = 0 \tag{140}$$

$$C_{ABCD}^{RSTU} = d_{AB}^{RS} d_{CD}^{TU} - d_{AC}^{RS} d_{BD}^{TU} + 16 \text{ further permuted terms} \tag{141}$$

Define now the projection operators $\mathsf{P}_0$ and $(1 - \mathsf{P}_0)$. Let $\mathsf{P}_0$ be a projector to the space spanned by $|\Phi_0\rangle$; $(1 - \mathsf{P}_0)$ is then the projector to the complementary space. Equation (128) implicitly contains the equation for the correlation energy (132) (given in terms of $\mathsf{t}_2$ operators) as well as the corresponding nonlinear equations for these operators. The whole problem of obtaining the correlation energy is that of isolating the mentioned equations. Equation (132), which is, in fact, the Schrödinger equation modified by the second quantization can be understood as a projection of Eq. (128) by $\mathsf{P}_0$ while Eqs. (137) and (138) are projections into the space $(1 - \mathsf{P}_0)$. Of course, since we are using $\mathsf{T} \approx \mathsf{T}_2$, our space $(1 - \mathsf{P}_0)$ is spanned by doubly excited Slater determinants $|\Phi_{AB}^{RS}\rangle$ only. On applying these projections to

$$\mathsf{H}\, e^{\mathsf{T}_2} |\Phi_0\rangle = E\, e^{\mathsf{T}_2} |\Phi_0\rangle \tag{142}$$

we get

$$\begin{aligned}\langle\Phi_0|\mathsf{H}\, e^{\mathsf{T}_2}|\Phi_0\rangle &= \langle\Phi_0|\mathsf{H}(1 + \mathsf{T}_2)|\Phi_0\rangle \\ &= E\langle\Phi_0|e^{\mathsf{T}_2}\Phi_0\rangle = E\, d_{AB}^{RS}\end{aligned} \tag{143}$$

and

$$\begin{aligned}\langle\Phi_{AB}^{RS}|\mathsf{H}\, e^{\mathsf{T}_2}|\Phi_0\rangle &= \langle\Phi_{AB}^{RS}|\mathsf{H}\left(1 + \mathsf{T}_2 + \frac{\mathsf{T}_2^2}{2}\right)|\Phi_0\rangle \\ &= E\langle\Phi_{AB}^{RS}|e^{\mathsf{T}_2}\Phi_0\rangle = E\, d_{AB}^{RS}\end{aligned} \tag{144}$$

Substituting for $E$ in Eq. (144) with (143) we get:

$$\langle\Phi_{AB}^{RS}|\mathsf{H}\left(1 + \mathsf{T}_2 + \frac{1}{2}\mathsf{T}_2^2\right)|\Phi_0\rangle = \langle\Phi_0|\mathsf{H}(1 + \mathsf{T}_2)|\Phi_0\rangle\, d_{AB}^{RS} \tag{145}$$

Expressing this in terms of matrix elements we obtain CPMET equations

$$\langle\Phi_{AB}^{RS}|\mathsf{H}|\Phi_0\rangle + \sum_{C<D}\sum_{T<U}\langle\Phi_{AB}^{RS}|\mathsf{H}|\Phi_{CD}^{TU}\rangle d_{CD}^{TU}$$

$$+ \frac{1}{2}\sum_{\substack{C<D \\ C,D\neq A,B}}\sum_{\substack{T<U \\ T,U\neq R,S}}\{(d_{AB}^{RS}d_{CD}^{TU} - d_{AC}^{RS}d_{AB}^{TU} + \ldots) + (d_{CD}^{TU}d_{BD}^{RS} - d_{BD}^{TU}d_{AC}^{RS} + \ldots)\}$$

$$\times \langle \Phi_{AB}^{RS} | H | \Phi_{ABCD}^{RSTU} \rangle = \langle \Phi_0 | H | \Phi_0 \rangle d_{AB}^{RS} + \sum_{C<D} \sum_{T<U} \langle \Phi_0 | H | \Phi_{CD}^{TU} \rangle$$

$$\times d_{CD}^{TU} d_{AB}^{RS} \tag{146}$$

It is possible to arrive from Eq. (145) at Eqs. (137) and (138) by realizing that

$$d_{AB}^{RS} = \langle RS | t_2 | AB \rangle - \langle RS | t_2 | BA \rangle \tag{147}$$

and by expressing the matrix elements in (146) over the orbitals[52].

# VI. Computational Methods and Results

In this section a practical point of view is emphasized. This means that we disregard methods which, in our opinion, are not perspective for one reason or another. The drawback of the Hylleraas method is its applicability only to the two-electron problem. We also omit here the methods based on the concept of separated electron pairs (geminals)[54–57] because these methods are inherently incapable of accounting for the interpair correlation energy[e]. We shall also not discuss here the Bethe-Goldstone equations since, from the calculations reported[7, 58] for BH and $H_2O$, it appears that they are computationally not suitable for chemical applications.

## A. Empirical Methods

Let us start a survey of computational methods by beginning with the simplest conceivable approach[59], that of assuming a constant empirical value for each type of bond and therefore obtaining the total correlation energy by summing the bond contributions (similarly in the same way that estimates are made in chemistry, *e.g.* for the heats of formation and dipole moments). We think that this method can hardly be refined to a state that would approach the so-called "chemical accuracy" because it disregards the interpair electron correlation.

Empirical parameters are involved in the EPCE-F2$\sigma$ method of Sinanoğlu and Pamuk[60, 61] as well as the method suggested by Clementi and co-workers[62, 63]. The two methods may be called semiempirical because they are based on the formulas given by the theory but adopt approximations involving empirical parameters. For the sake of compactness, the EPCE-F2$\sigma$ method will be discussed in Section VI.B. dealing with Sinanoğlu's Many-Electron Theory. The formula in the method of

---

e) A single antisymmetrized product of strongly orthogonal geminals is incapable of accouting for interpair correlation. Since the last decade, however, this term is used as a synonym for the method of expansion of wave functions in terms of orthogonal localized orbitals. The geminal product wave function serves only as a tool to determine a localized orthonormal basis, *i.e.* "excited" orbitals, to describe correlation which can be applied to construct the cluster-functions. (E. Kapuy: private communication.)

Clementi and co-workers is a functional which is a modified expression of Wigner's[64] derived for the electron gas. Mathematically, it is an integral containing powers of the electron density. If the Hartree-Fock density is used, the use of the method is restricted to cases where the Hartree-Fock function is a good representation of the exact wave function. For example, for dissociation processes the electron density based on a proper (CI) wave function must be employed. Examples of applications for which the use of the Hartree-Fock electron density seems to be sufficient are the water dimer[65] and the complexes[66] of the water molecule with $Li^+$, $Na^+$, $K^+$, $F^-$ and $Cl^-$ ions. The computed correlation energies for these interactions are considerably underestimated, however. The numerical integration of the density functional is claimed[67] to be much shorter than the computation of integrals in the Hartree-Fock calculation. As regards the accuracy and reliability of the method we cite the authors[68]: "Whether the data here obtained appear to be "good" or "bad" depends on one's purpose. For us the data are preliminary but indicative that the search for a proper reference function and a more versatile functional of the density is a task worth pursuing".

## B. Many-Electron Theory (MET) of Sinanoğlu

Sinanoğlu[47] was the first to introduce to quantum chemistry methods based on the cluster expansion. We are not going to present a derivation of MET here. We only note that one can arrive at the MET equations using Čížek's CPMET, by neglecting all terms which couple $T_2$ matrix elements corresponding to different hole pairs. In this manner, the CPMET equations are reduced to independent equations for each pair of the occupied orbitals. It is therefore possible[50] to refer to MET as to the decoupled-pair-many-electron theory. In Sinanoğlu's MET as in CPMET, the approximation $T \approx T_2$ is assumed. For the function

$$|\Psi\rangle = |\Phi_0\rangle + \sum_{i<j} t^{(ij)} |\Phi_0\rangle \tag{148}$$

Sinanoğlu derived the expression

$$E \approx E_{HF} + \frac{1}{D} \sum_{i<j} \tilde{\epsilon}_{ij} \tag{149}$$

in which $\tilde{\epsilon}_{ij}$ are pair correlation energies and $D$ is related to the normalization of $|\Psi\rangle$. Pair correlations are determined from the so-called pair functions (in our notation these are clusters $t^{(ij)}|\Phi_0\rangle$) which are obtained independently for each pair of occupied spin-orbitals. In this simplest form, MET is equivalent to IEPA (see Section VI.C.). If the unlinked clusters are included in an approximation, Sinanoğlu can transform the function

$$|\Psi\rangle = |\Phi_0\rangle + \sum_{i<j} t^{(ij)} |\Phi_0\rangle + \sum_{\substack{i<j \\ i,j \neq k,l}} \sum_{k<l} t^{(ij)} t^{(kl)} |\Phi_0\rangle \tag{150}$$

into the following expression

$$E \approx E_{\mathrm{HF}} + \sum_{i<j} \tilde{\epsilon}_{ij} \frac{D_{ij}}{D'} \tag{151}$$

This equation and (149) differ only by the normalization factors. For $N \to \infty$ it was shown[47] that $(D_{ij}/D') \to 1$, so that it is possible to write

$$E \approx E_{\mathrm{HF}} + \sum_{i<j} \tilde{\epsilon}_{ij} \tag{152}$$

The most important result of the papers by Sinanoğlu on MET was the finding that in a CI treatment with quadruply excited configurations, the linked tetra-excited part $(T_4)$ is negligible in comparison to the unlinked part $\left(\frac{1}{2} T_2^2\right)$. This was clearly shown by Sinanoğlu in an analysis of the CI calculation performed by Watson[69] for the beryllium atom. Watson's CI wave function contained the 37 most important configurations. Among them were four quadruply excited configurations whose CI expansion coefficients are given in Table 2. The entries in the last column of Table 2

Table 2. Four-electron correlation and unlinked clusters in Be atom[47]

| Quadruply excited configuration[1]) | Coefficient from 37-configuration wave function[69]) | Coefficient calculated from double excitations by Eq. (141) |
|---|---|---|
| $p_{\mathrm{I}}^2 p_{\mathrm{II}}^2$ | 0.007063 | 0.0073 |
| $p_{\mathrm{I}}^2 s_{\mathrm{I}}^2$ | 0.005651 | 0.00647 |
| $p_{\mathrm{I}}^2 d_{\mathrm{II}}^2$ | 0.001585 | 0.00168 |
| $p_{\mathrm{II}}^2 d_{\mathrm{I}}^2$ | 0.000464 | 0.000478 |
| Energy contribution | −0.075 eV | −0.074 eV |

[1]) Roman numerals denote the order of virtual orbitals.

are coefficients given for the products of the expansion coefficients of the doubly excited respective configurations. For the configurations $p_{\mathrm{I}}^2 s_{\mathrm{I}}^2$, for example, the relevant doubly excited configurations involved in Watson's wave function have the following expansion coefficients: $1s^2 p_{\mathrm{I}}^2$ – 0.2844586; $2s^2 s_{\mathrm{I}}^2$ – 0.0232595; $2s^2 p_{\mathrm{I}}^2$ – 0.001899; $1s2sp_{\mathrm{I}}^2$ 0.0065906 und $1s2ss_{\mathrm{I}}^2$ – 0.0055013. Denoting these coefficients by brackets, according to Eq. (141) we get

$$C_{1s^2 2s^2}^{p_{\mathrm{I}}^2 s_{\mathrm{I}}^2} \text{ (unlinked)} = [1s^2 p_{\mathrm{I}}^2][2s^2 s_{\mathrm{I}}^2] + 4[1s2sp_{\mathrm{I}}^2][1s2ss_{\mathrm{I}}^2] = 0.00647$$

From Table 2 one can see that the data given by the approximation $C_4 \approx \frac{1}{2} T_2^2$ are in good agreement with the results of the complete CI treatment and that the effect of linked $T_4$ clusters is therefore very small.

Sinanoğlu also derived the LCAO form of MET and suggested a series of semi-empirical procedures for estimating the correlation energy. Among them the one in most general use is the so-called EPCE-F2$\sigma$ method formulated by Sinanoğlu and Pamuk[60, 61]. This method is based on Eq. (152), which in the LCAO form becomes

$$E_{\mathrm{corr}} \approx \frac{1}{4} \sum_{p,q} P_{pp} P_{qq} \bar{\epsilon}_{pq} \qquad (153)$$

where $P_{ii}$ are diagonal elements of the bond order matrix and

$$\bar{\epsilon}_{pq} \equiv \epsilon_{p\alpha q\beta} + \epsilon_{p\alpha q\alpha} \qquad (154)$$

is the effective pair correlation energy (EPCE) for which we assume the so-called "F2 approximation":

$$\epsilon_{p\alpha q\beta} \approx 2\, \epsilon_{p\alpha q\alpha} \qquad (155)$$

The last symbol in the designation of the method means that, in contrast to the previous version of the method[70], it is not restricted to $\pi$-electronic systems.

Computationally, one first performs a semiempirical all-valence electron calculation (say CNDO/2) to obtain the electron densities $P_{ii}$. One-center $\bar{\epsilon}_{pq}$ are constants which were tabulated[60] for H, B, C, N, O and F atoms. Two-center $\bar{\epsilon}_{pq}$ are evaluated by an empirical formula which is a function of the interatomic distance. Computation of the expression (153) is much shorter than a standard CNDO/2 run. It is understandable that such a simple method cannot provide highly accurate estimates of the correlation energy. Actually, this accuracy is claimed[61] to be within ±0.5 eV.

Table 3. Interaction correlation energy (in kJ/mole) in van der Waals complexes[75]

| Complex[1] | EPCE-F2$\sigma$ | IEPA intersystem $\Delta E_{\mathrm{corr}}$ |
|---|---|---|
| 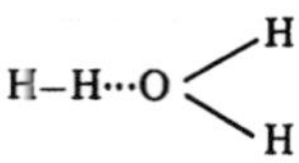 | −2.798 | −3.578[2] |
| 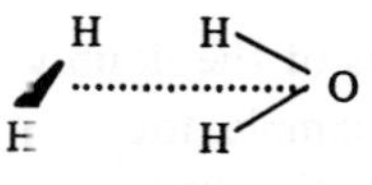 | −2.605 | −1.580[2] |
| H–H···F–H | −0.965 | −1.268[2] |
| H, H (H–H)···H–F | −11.385 | −4.127[2] |
| H–F···H–F | −4.921 | −3.366[3] |

[1] The geometries assumed refer to the van der Waals minima given by IEPA calculations.
[2] Ref.[87].
[3] Ref.[88].

Although this tolerance is much larger than that required in many chemical problems, the feasibility of calculations warrants attempts at application. These were attempted for binding energies[61)], ionization potentials[71, 72)], excitation energies[72)], clusters of hydrogen atoms[73)] and intermolecular interactions[74, 75)]. In Table 3 we present some results for van der Waals complexes.

## C. Independent Electron Pair Approximation (IEPA)

This approximation has been known for a long time, *e.g.* in the form of Sinanoğlu's MET or Nesbet's formulation of the second-order Bethe-Goldstone method[76)]. Practical calculations with the IEPA method, however, were first developed by Kutzelnigg's group[77, 78)]. In the language of configuration interaction, one uses the wave function

$$|\Psi_{AB}\rangle = |\Phi_0\rangle + \sum_{R<S} C_{AB}^{RS} |\Phi_{AB}^{RS}\rangle \tag{156}$$

This means that we perform a CI calculation, separately, for each pair of occupied spin-orbitals $A, B$. This gives us an energy increment $\epsilon_{AB}$ which can be ascribed to the electron correlation in spin-orbitals $A, B$. Since the electron pairs are considered to be independent, the total correlation energy is given by

$$E_{\text{corr}} = \sum_{A,B} \epsilon_{AB} \tag{157}$$

For technical details of calculations such as the use of pair natural orbitals, we refer the reader to the cited papers[77, 78)]. Here we restrict ourselves to a few comments on the properties of IEPA correlation energy. IEPA is not a variational method and it does not give an upper bound to the energy. Actually, the correlation energy is overestimated, in some cases by as much as 30%. The situation is somewhat better if localized orbitals are used, but the estimated correlation energies are still too high. The extent of this overestimation depends on the particular case and it is, therefore, possible to speak about "good" and "poor" IEPA molecules[79)]. An example of the failure of IEPA is presented for the $F_2$ dissociation in Section VI.I. Among the advantages of IEPA, the major one is the computational economy. This allows one to undertake systematic studies of the effect of correlation energy on molecular geometries and force constants[32, 80–83)], potential curves[79, 83)], and van der Waals interactions[84–88)]. Another advantage of IEPA is the correct dependence[3, 78)] of the correlation energy to the number of electrons, *i.e.* the energy of a supersystem of $n$ noninteracting subsystems, $E(nA)$, is equal to the sum of the energies of the subsystems, $nE(A)$.

## D. Modified Versions of the Configuration Interaction Method

As stated in the introduction we are not going to pay much attention to the CI method because there are numerous reviews on this topic (*e.g.* [10–13)]). We only briefly note some techniques that treat the CI expansion in a more effective way. We also note the properties of the CI wave function which have some relation to other methods considered in this review.

The traditional CI treatment which is nowadays commonly referred to as "brute force" CI, is inpractical for routine calculations of correlation energy because of the slow energy convergence of the CI expansion. For the purpose of comparison with other methods, it is advantageous to adopt the traditional CI. What we need to know, is the role of the singly, doubly, triply, *etc.* excited configurations in the CI expansion. We limit ourselves to the most common case – the Slater determinant of a closed shell ground state formed from the Hartree-Fock orbitals. In this case, the effect of singly excited configurations is small, as the consequence of Brillouin's theorem. However, it appears that in some special applications (other than energy predictions) single excitations can not be omitted in the CI expansion. For example, singly excited states are responsible for the correct sign of the dipole moment of CO[89]. The main contribution (more than 90%) to the correlation energy comes from the doubly excited states. For larger than diatomic molecules, it is difficult, for practical reasons, to test the effect of triple and higher excitations by way of direct calculations. Therefore, it is only possible to do speculative estimates on the basis of data for atoms and diatomic molecules. For example, in the case of the Be atom[90] the triple excitations correspond to 0.3% of the calculated correlation energy and the quadruple excitations to 3.8%. In $N_2$ the contribution[91] of quadruply excited configurations is 7%. It may be assumed, therefore, that for small polyatomic molecules it is sufficient to consider only doubly excited configurations and to also include singly and quadruply excited configurations when striving for highly accurate calculations. This is in contrast to the opinion of Davidson[11] who states that for a large molecule, a hydrocarbon chain longer than about 30 carbons, most of the correlation energy should come from the quadruple or higher excitations.

In spite of the truncation of the CI expansion so as to include the double excitations, the ordinary CI calculations are hardly more practical. There are many devices and tricks[10–13] which facilitate the problem: use of symmetry, selection of configurations by perturbation estimates of their weights, efficient diagonalization algorithms, economic computer handling of large CI-matrices[92], avoiding the explicit construction of the CI-matrix[93], effective transformations of integrals from AO to MO basis, disregarding configurations that contribute to inner shell correlation energy, energy extrapolations[94] and others. These are, however, of more or less technical nature and do not remove the inherent drawback of the ordinary CI expansion – this being its slow energy convergence. A very efficient way of obtaining a more rapidly convergent CI expansion is based on the use of natural orbitals. We take note of two techniques of this type at this time. In the Iterative Natural Orbital method (INO-CI) developed by Bender and Davidson[95], one proceeds[5,12] as follows: A certain number of configurations, say 50, are selected and a CI calculation is carried out. The density matrix given by the CI wave function is diagonalized which gives us the natural orbitals. From these the same set of configurations is constructed and the CI calculation is repeated. Several iterations are performed according to this scheme until the energy reaches a minimum. An important feature of the method is that unimportant configurations are deleted in the iteration process and new configurations are added. The choice of singly and doubly excited configurations may be made in an ingenious way (*e.g.* for $H_2O$[96] and $NH_2$[97] that permits consideration of the valence-shell correlation energy. Some applications of the method are presented in the book by Schaefer[5]. The second method we are going to note is perhaps more economic. It is the so called Pair Natural Orbital method (PNO-CI) developed by Meyer[98]. PNO-CI may be taken as an extension of the IEPA-PNO method. As in the IEPA-PNO one calculates the pair natural orbitals for each pair of occupied spin-orbitals and constructs from them the (lowest) doubly excited states. But in contrast to IEPA-PNO, the PNO-CI wave functions contains $|\Phi_0\rangle$ and doubly excited configurations, which correspond to excitations from all pairs. This, of course, brings about difficulties in constructing the Hamiltonian matrix elements, $H_{ij}$, because each pair generates its own set of PNO's. In other words, PNO's for the pair $R, S$ are not orthogonal to PNO's for the pair $T, U$. Fortunately, the nonorthogonality does not represent a serious problem. Although the CI-matrix $H_{ij}$ elements[78] are somewhat more complex than those given by Slater rules for an orthogonal set, they are still tractable. It should be emphasized that the same $H_{ij}$ elements appear in the CEPA-PNO equations. It is therefore-profitable to perform PNO-CI and CEPA-PNO calculations in a single run.

Let us now comment briefly on the general properties of the CI wave function with only singly and doubly excited configurations (it is not of importance whether it was obtained by "brute force", INO-CI or the PNO-CI treatment). As we already know it should cover a large portion of the correlation energy. Another advantage is that it furnishes an upper bound to the energy because CI is a genuine variational method. A drawback of this limited CI wave function is the incorrect dependence on the number of particles and its incorrect behaviour in dissociation (see Section VI.I.). To illustrate the incorrect dependence on the number of particles we make use of the PNO-CI data of Ahlrichs[19] for the dimerization of $BH_3$. He arrived at the following energies of dimerizations

$$\Delta E = E(B_2H_6) - 2\,E(BH_3) = -27.4 \text{ kcal/mole} \tag{158}$$

$$\Delta E = E(B_2H_6) - E(2\,BH_3) = -34.2 \text{ kcal/mole} \tag{159}$$

In Eq. (158) twice the PNO-CI energy of $BH_3$ was assumed, whereas in Eq. (159) the two $BH_3$ molecules were represented by a supersystem $BH_3BH_3$ at large intermolecular distance (50 a.u.). In the first case, the two $BH_3$ molecules are strongly biased with respect to $B_2H_6$. In fact, we consider simultaneous double excitation of either $BH_3$, the net result of which means inclusion of quadruply excited configurations. In order to obtain a balanced description of both sides of the reaction, it is necessary to treat it by means of Eq. (159). This trick may be applied to dissociations where the SCF function behaves correctly at the dissociation limit. If it does not, the inclusion of quadruply excited configurations cannot be avoided and the CI treatment, limited to doubly excited configurations, cannot be applied (see Section VI.I.).

With respect to the multiconfiguration SCF (MCSCF), the virtues and the drawbacks of this method[99–101] have been described in the literature on several occasions (see *e.g.* Ref.[5]). In our opinion, the MCSCF still remains much more suited for the solution of near-degeneracy problems such as the removal of discontinuities, cusps, humps and other artefacts in the SCF energy hypersurfaces as reported by Gregory and Paddon-Row[102, 103], than to be used in large scale calculations including correlation energy.

### E. Coupled-Electron Pair Approximation (CEPA)

CEPA[98, 104] represents one of the most successful approaches to the calculation of the correlation energy of molecules from the viewpoint of accuracy and the expenditure of computer time. CEPA may be looked upon as an approximation of CPMET. Again the wave function is assumed to have the form

$$|\Psi\rangle = e^{T_2}|\Phi_0\rangle \tag{160}$$

CEPA is based on the approximation of the third term on the left hand side of the Eq. (146). We assume

$$\begin{aligned}&\frac{1}{2}\sum_{\substack{C<D\\C,D\neq A,B}}\sum_{\substack{T<U\\T,U\neq R,S}}\{(d_{AB}^{RS}d_{CD}^{TU} - d_{AC}^{RS}d_{BD}^{TU} + \ldots)\\&+ (d_{CD}^{TU}d_{AB}^{RS} - d_{BD}^{TU}d_{AC}^{RS} + \ldots)\} < \Phi_{AB}^{RS}|\mathrm{H}|\Phi_{ABCD}^{RSTU}\rangle\\&\approx d_{AB}^{RS}\sum_{C,D\neq A,B}\sum_{T<U}\langle\Phi_{CD}^{TU}|\mathrm{H}|\Phi_0\rangle d_{CD}^{TU}\end{aligned} \tag{161}$$

Note that the terms in parentheses on the left hand side of Eq. (161) contain products of $d$ coefficients and represent the unlinked part of the CI expansion coefficient for the quadruply excited configuration $|\Phi_{ABCD}^{RSTU}\rangle$ [compare Eq. (141)]. Within the CEPA these terms are now substituted for by only a single product of $d$'s. This approximation was suggested by Kelly[44, 105] but was first employed for practical calculations by Meyer. The justifiability of this approximation was shown by time-dependent diagrams[7, 105]. We attempted to present this approximation by making use of the language of diagrams in the sections describing the MB-RSPT and the cluster expansion method.

Let us now rewrite the term which we are going to approximate in terms of $t_2$ operators. First we express it as

$$\langle\Phi_{AB}^{RS}|\mathrm{H}|\ \frac{1}{2}\mathrm{T}_2^2\,\Phi_0\rangle \tag{162}$$

As we know from Section III, the Slater determinant for a doubly excited configuration can be written as

$$|\Phi_{AB}^{RS}\rangle = N[\mathrm{X}_R^+\mathrm{X}_S^+\mathrm{X}_A\mathrm{X}_B]|\Phi_0\rangle \tag{163}$$

and the term (162) becomes

$$\langle \Phi_0 | N[X_B^+ X_A^+ X_S X_R] \{H\} \frac{1}{2} T_2^2 | \Phi_0 \rangle \tag{164}$$

According to Eq. (37) we know that in expression (164) all the operators must be contracted. Therefore, if we substitute the normal product form (48) for H, we have to consider only the two-electron part of the Hamiltonian, *i.e.* we can assume v instead of H in the expression (164). Hence, we have

$$\langle \Phi_0 | N[X_B^+ X_A^+ X_S X_R] \left\{ \frac{1}{2} \sum_{EFGH} \langle EF | v | GH \rangle N[X_E^+ X_F^+ X_H X_G] \right\}$$
$$\times \left\{ \frac{1}{2} \sum_{CDTU} \sum_{IJWZ} \langle CD | t_2 | TU \rangle \langle IJ | t_2 | WZ \rangle N[X_C^+ X_D^+ X_U X_T] \right.$$
$$\left. \times N[X_I^+ X_J^+ X_Z X_W] \right\} | \Phi_0 \rangle \tag{165}$$

This can be schematically represented by diagrams as shown in Fig. 13. The uppermost left diagram in Fig. 13 represents the doubly excited configuration. Note that

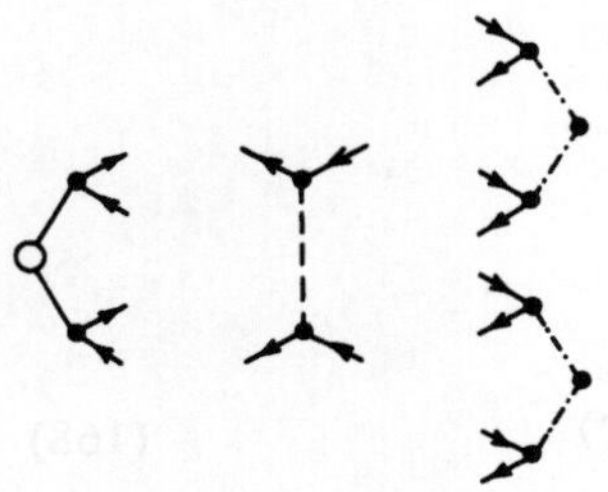

Fig. 13

there is no summation for *A, B, R, S* indices and it is this property which will select only the *A, B, R, S* indices in the summations referring to $t_2$ operators (by way of Kronecker's delta, which originates from contraction).

In order to obtain the proper mathematical expression for Fig. 13 we have to construct all necessary diagrams. In Fig. 14 we present characteristic types of diagrams which refer to different types of contributions. The latter being

upper left:

$$\sum_{\substack{C<D \\ C,D \neq A,B}} \sum_{\substack{T<U \\ T,U \neq R,S}} \langle CD | v | TU \rangle \langle TU | t_2 | CD \rangle \langle RS | t_2 | AB \rangle \tag{166}$$

upper right:

$$\sum_{\substack{C<D \\ C,D \neq A,B}} \sum_{\substack{T<U \\ T,U \neq R,S}} \langle CD | v | TU \rangle \langle SU | t_2 | CD \rangle \langle RT | t_2 | AB \rangle \tag{167}$$

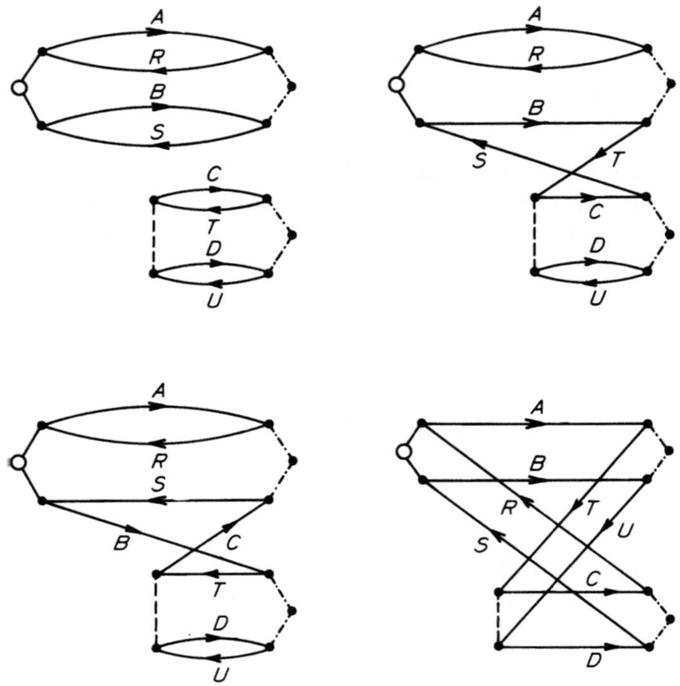


Fig. 14. Characteristic types of diagrams for the expression (165)

lower left:

$$\sum_{\substack{C<D\\ C,D\neq A,B}} \sum_{\substack{T<U\\ T,U\neq R,S}} \langle CD|\mathrm{v}|TU\rangle\langle TU|\mathrm{t}_2|BD\rangle\langle RS|\mathrm{t}_2|AC\rangle \tag{168}$$

lower right:

$$\sum_{\substack{C<D\\ C,D\neq A,B}} \sum_{\substack{T<U\\ T,U\neq R,S}} \langle CD|\mathrm{v}|TU\rangle\langle RS|\mathrm{t}_2|CD\rangle\langle TU|\mathrm{t}_2|AB\rangle \tag{169}$$

Assume that the spin-orbitals are well localized and that the spin-orbital pair $A, B$ is well separated from the spin-orbital pair $C, D$. Then it may also be assumed that the virtual spin-orbitals are located in different spaces and the latter may be assigned to spaces of the occupied spin-orbitals. Thus, if $\langle CD|\mathrm{v}|TU\rangle$ should be large then $T$ and $U$ must correspond to $C$ and $D$. Under this assumption, at least one of the two $\mathrm{t}_2$-matrix elements appearing in each of the expressions (167)–(169) can be expected to be small. Therefore, the contributions (167)–(169) may be expected to be much smaller than the contribution (166). If the former are neglected and the latter is completed to include its exchange terms, the expression (162), by making use of Eq. (147), becomes

$$\left\langle \Phi_{AB}^{RS} | \mathrm{H} | \frac{1}{2} \mathrm{T}_2^2 \Phi_0 \right\rangle = \sum_{\substack{C<D \\ C,D \neq A,B}} \sum_{\substack{T<U \\ T,U \neq R,S}} \{\langle CD | v | TU \rangle - \langle CD | v | UT \rangle\} d_{CD}^{TU} d_{AB}^{RS} \tag{170}$$

We rewrite Eq. (170) by relaxing condition $T, U \neq R, S$ and substract the new terms in the second summation. We thereby obtain

$$\begin{aligned} \left\langle \Phi_{AB}^{RS} | \mathrm{H} | \frac{1}{2} \mathrm{T}_2^2 \Phi_0 \right\rangle = & \sum_{\substack{C<D \\ C,D \neq A,B}} \sum_{T<U} \{\langle CD | v | TU \rangle - \langle CD | v | UT \rangle\} \\ & \times d_{CD}^{TU} d_{AB}^{RS} - \sum_{\substack{C<D \\ C,D \neq A,B}} \left\{ \sum_T [(\langle CD | v | RT \rangle - \langle CD | v | TR \rangle) \right. \\ & \times d_{CD}^{RT} + (\langle CD | v | ST \rangle - \langle CD | v | TS \rangle) d_{CD}^{ST}] \\ & \left. + (\langle CD | v | RS \rangle - \langle CD | v | SR \rangle) d_{CD}^{RS} \right\} d_{AB}^{RS} \end{aligned} \tag{171}$$

By using the same arguments as in the discussion for the expressions (166)–(169) we can neglect the second term in Eq. (171) and we obtain the following equation

$$\left\langle \Phi_{AB}^{RS} | \mathrm{H} | \frac{1}{2} \mathrm{T}_2^2 \Phi_0 \right\rangle = \sum_{\substack{C<D \\ C,D \neq A,B}} \sum_{T<U} \{\langle CD | v | TU \rangle - \langle CD | v | UT \rangle\} \times d_{CD}^{TU} d_{AB}^{RS} \tag{172}$$

which is equivalent to Eq. (161).

We can now define the pair correlation energies $\epsilon_{CD}$ as

$$\epsilon_{CD} = \sum_{T<U} \{\langle CD | v | TU \rangle - \langle CD | v | UT \rangle\} d_{CD}^{TU} \tag{173}$$

From inspection of Eq. (143) it follows that

$$E - \langle \Phi_0 | \mathrm{H} | \Phi_0 \rangle = \sum_{C<D} \sum_{T<U} \langle \Phi_0 | \mathrm{H} | \Phi_{CD}^{TU} \rangle d_{CD}^{TU} = \sum_{C<D} \epsilon_{CD} \tag{174}$$

Equation (172) can now be written in a very compact form:

$$\left\langle \Phi_{AB}^{RS} | \mathrm{H} | \frac{1}{2} \mathrm{T}_2^2 \Phi_0 \right\rangle = \sum_{\substack{C<D \\ C,D \neq A,B}} \epsilon_{CD} d_{AB}^{RS} \tag{175}$$

Introducing the approximation (175) into the CPMET equations (146) we obtain the CEPA equations

$$\begin{aligned} \langle \Phi_{AB}^{RS} | \mathrm{H} | \Phi_0 \rangle + \sum_{C<D} \sum_{T<U} \langle \Phi_{AB}^{RS} | \mathrm{H} | \Phi_{CD}^{TU} \rangle d_{CD}^{TU} = \\ = d_{AB}^{RS} \langle \Phi_0 | \mathrm{H} | \Phi_0 \rangle + d_{AB}^{RS} \epsilon_{AB} \end{aligned} \tag{176}$$

The CEPA Eqs. (176) are solved iteratively. One may start with the IEPA pair correlation energies, for example, and obtain the *d* coefficients from Eq. (176). These can then be used to evaluate new pair correlation energies using Eq. (173) which then can be used in Eq. (176) in the next step. Alternatively[98, 106], instead of Eq. (176) one can solve the CI problem by shifting the diagonal elements by the sum of the energy contributions due to other pairs, *i.e.* the $D_{AB}$ elements are replaced by $D'_{AB}$ terms, their meaning being

$$D_{AB} = \langle \Phi_{AB}^{RS} | \mathrm{H} | \Phi_{AB}^{RS} \rangle \tag{177}$$

and

$$D'_{AB} = \langle \Phi_{AB}^{RS} | \mathrm{H} | \Phi_{AB}^{RS} \rangle + \sum_{C,D \neq A,B} \epsilon_{CD} \tag{178}$$

However the first approach is preferable for computational reasons.

The CEPA computer programs, developed by two German groups[78, 104], are based on the use of pair natural orbitals and, therefore, it is appropriate to refer to their approach as the CEPA-PNO method. The construction of **H**-matrix elements over the PNO's is the same as with the PNO-CI (this was noted in the preceding section).

CEPA was tested systematically on a series of small molecules[32, 79, 106–109]. Excellent agreement with experiment was found for molecular geometries, spectroscopic constants, dipole moments, dissociation energies, ionization potentials and electron affinities. As regards the potential curves, CEPA gives very good agreement with experiment over a relatively large region around the equilibrium distance. At larger distances the CEPA energy starts to deviate from the experimental curve and it does not converge towards the correct dissociation limit. This is due to the fact that the approximation inherent in the CEPA treatment of unlinked clusters is not well satisfied in this region (see Section VI.I.).

CEPA is not a variational method, but the high accuracy achieved in the calculated molecular properties suggests that the advantage of furnishing an upper bound to the energy is not so important. In contrast to PNO-CI, the inclusion of unlinked clusters ensures the correct dependence with respect to the number of particles. As noted above in the derivation of CEPA equations, the basic approximation (161) of CEPA is justifiable for well separated pairs. If this condition is not satisfied, the effect of quadruple excitations is most likely overestimated[104], although the effect of the approximation (161) is not yet well established. The portion of the correlation energy accounted for, depends on the size of the basis set. Polarization functions were found to be very important because they contribute much more to the correlation energy than to the SCF energy[32, 109]. Roughly speaking, CEPA gives about 85% of the total correlation energy for basis sets containing two sets of polarization functions[32]. To illustrate the computer time required for such a calculation we present the data reported by Ahlrichs *et al.*[32] in Table 4. For the sake of completeness, we take note of the Independent Pair-Potential Approximation (IPPA) by Mehler[110]. Using the flow-chart of the CEPA method he modified the IEPA method by including a class of coupling elements.

Table 4. CEPA calculation[32)] of planar $NH_3$ with 172 Gaussian lobes in 58 groups on a UNIVAC 1108

| Operation[1)] | CPU time (in min.) |
|---|---|
| Evaluation of the integrals | 73 |
| Hartree-Fock (7 iterations) | 4 |
| Construction of operators[2)] | 12 |
| Calculation of PNO (11 pairs)[3)] | 12 |
| Construction of the diagonal CI blocks | 76 |
| Construction of the off-diagonal blocks | 4 |
| Evaluation of $E_{CEPA}$ and $E_{CI}$[4)] | 2 |
| Miscellaneous | 4 |
| Total | 187 |

1) For details see Ref.[78)].
2) Coulomb $J^R$, $J^{RS}$ and exchange $K^R$, $K^{RS}$ operators.
3) The other five pairs are obtained by reflection.
4) CEPA-PNO and PNO-CI calculations are closely related and they are performed in a single run.

## F. Perturbation Calculations

This section focuses on two practical points: the portion of the correlation energy covered by the perturbation expansion and the cost involved. We are not going to attempt to give a complete bibliography of papers dealing with MB-RSPT calculations of the correlation energy. Instead, we shall mention only those that are relevant to the problem noted, as well as, those that are closely related to the computational scheme outlined in Section IV.

Bartlett and Silver[111)] used MB-RSPT up to the third order with a denominator shift. Actually, their approach should be called a "two-body" PT, because the correlation energy was calculated as the sum of contributions of individual pairs of occupied spin-orbitals. The molecules treated were LiH, BH and HF. With the largest of the Slater basis sets used, the calculations recovered 94, 95 and 97% of the experimental correlation energy, respectively. Using the same approach, Bartlett and Silver obtained[112)] about 80% of the correlation energy of the NaH, AlH and HCl molecules. As they demonstrated, the pair restriction, as was the case with the IEPA method, gives correlation energies that are too high, the overestimation being typically 10–15%. They also tested the various types of one-particle potentials such as $V^N$ and $V^{N-1}$. The effect was found negligible when all second and third order diagrams were taken into account[113)]. Freemen and Karplus[114)] used MB-RSPT to calculate the correlation energy of $Li_2$, $N_2$ and $H_3$ molecules. They used the hole-line expansion to determine the diagrams which contribute to various orders of MB-RSPT in comparison to the CI method. With very small basis they obtained approximately 72% of the correlation energy compared with the full CI treatment. They found that for a small basis the MB-RSPT converges rather slowly, whereas, with an extended basis set the convergence appears to be better. For the $N_2$ molecule they used a very large basis and compared the second order of the MB-RSPT, as well as, the second-order Epstein-Nesbet partitioning of the Hamiltonian with the experimental correlation energy. From their results one can see that the second order of MB-RSPT gives approximately 95% of the experimental correlation energy and that the second-order Epstein-Nesbet partitioning overestimates it significantly.

A systematic study devoted to the MB-RSPT was undertaken by Pople *et al.*[115]. The utility of the theory was demonstrated by calculations up to the second and third orders on the equilibrium geometries, dissociation energies and energy differences between the electronic states of different multiplicity.

Comparison with experiment is certainly meaningful but, in our opinion, comparing the MB-RSPT results with the CI data for exactly the same basis set, thereby, allowing an estimate of the amount of correlation energy included in the second and third order of MB-RSPT, would also be highly useful. Since a systematic study of this kind is still lacking we performed[116] such a comparison for the $H_2O$, HF, Ne and BH systems. The correlation energies up to the third order were calculated by means of the formulas (82) and (84). For comparison, we decided to use the INO-CI results[96] for $H_2O$ including all singly and doubly excited configurations. From the five different basis sets used for the INO-CI calculations we have selected four contracted Gaussian sets, starting from the double zeta $[4s2p/2s]$ to double zeta plus polarization $[4s2p1d/2s1p]$ quality. The results are summarized in Table 5. In Table 6 we present results for BH, HF and Ne. We note that only valence shell correlation energies were calculated. No restrictions were imposed upon the number of used unoccupied orbitals given by the particular basis set. Our SCF energies for HF and Ne are slightly different from those reported by Ahlrichs *et al.*[32], (−100.04823 and −128.52516). This rather small deviation may perhaps be due to the fact that Ahlrichs *et al.* used Gaussian lobe functions, whereas, we used cartesian Gaussian functions. We believe that for our purposes this small discrepancy may be disregarded. Tables 5 and 6 show that the correlation energy of ten-electron systems included in $k_0^{(2)}$ and $k_0^{(3)}$ is about 100%, compared to CI-SD and CEPA calculations with the same basis set. As expected, the $k_0^{(2)}$ and $k_0^{(3)}$ contributions are slightly basis set dependent, the dependency being stronger for the latter than for the former. As is mentioned in Ref.[115], when one performs CI calculations truncated to include the singly plus the doubly excited configurations, the terms which behave incorrectly with $N^2$ are included in the CI energy. Therefore, one should look upon the CI-SD calculations with some caution. We note that the $k_0^{(2)}$ and $k_0^{(3)}$ of the MB-RSPT behave correctly in proportion to $N$.

The comparison of the second and the third order of the MB-RSPT and the CI-SD depends on two errors:

(i) the truncation of the CI expansion and
(ii) the truncation of the MB-RSPT series at the third order.

Specifically, tetra-excited states are responsible for the cancellation of the incorrect $N^2$ behaviour of the CI-SD, as well as, for the further reduction of the correlation energy. Due to truncation of the MB-RSPT at the third order the effect of doubly excited configurations are not included fully and, of course, the higher excited configurations are omitted completely.

Table 5. Valence shell correlation energy in $H_2O$[1]) given by MB-RSPT treatment[116]) and the INO-CI calculations[96]) including all singly and doubly excited configurations (CI-SD)

| Basis set[2]) | SCF | CI-SD | $k_0^{(2)}$ | $k_0^{(2)} + k_0^{(3)}$ | $\frac{k_0^{(2)} + k_0^{(3)}}{\text{CI-SD}}$ % |
|---|---|---|---|---|---|
| [4s2p/2s] | −76.009294 | −0.1257 | −0.1251 | −0.1262 | 100.4 |
| [4s2p/2s1p] | −76.033838 | −0.1393 | −0.1388 | −0.1408 | 101.1 |
| [4s2p1d/2s] | −76.036678 | −0.1839 | −0.1841 | −0.1892 | 102.9 |
| [4s2p1d/2s1p] | −76.048764 | −0.1930 | −0.1934 | −0.1990 | 103.1 |

1) Energies in a.u. The geometry assumed: $R_{OH} = 1.8089$ a.u., $\vartheta = 104.52°$.
2) For details see Ref.[96]).

Table 6. Comparison of valence shell correlation energies[1]) in BH, HF and Ne given by MB-RSPT[116]), CEPA and CI calculations including all singly and doubly excited configurations (CI-SD)

| System | SCF | CI-SD | CEPA | $k_0^{(2)}$ | $k_0^{(2)} + k_0^{(3)}$ | $\frac{k_0^{(2)} + k_0^{(3)}}{\text{CI-SD}}$ % |
|---|---|---|---|---|---|---|
| BH[2]) | −25.105638 | −0.0694[3]) | −0.0721 | −0.0477 | −0.0621 | 89.5 |
| HF[4,5]) | −100.048548 | −0.2171 | −0.2257 | −0.2263 | −0.2252 | 103.7 |
| Ne[4]) | −128.524067 | −0.2094 | −0.2149 | −0.2157 | −0.2145 | 102.4 |

1) Energies in a.u.
2) $R_{BH} = 2.336$ a.u.; [4s2p/2s1p] basis set used (for details see Ref.[117])): CI-SD and CEPA results from Ref.[117]).
3) The full-CI value[117]) is −0.0727.
4) [5s3p1d/3s1p] basis set used (for details see Ref.[32])); PNO-CI and CEPA-PNO results from Ref.[32]).
5) $R_{HF} = 1.733$ a.u.

Keeping this in mind, we shall now discuss the different behavior of molecules under study. For the BH molecule, which is a six-electron system, $k_0^{(2)}$ and $k_0^{(3)}$ includes about 90% of the correlation energy compared to the CI-SD results. Note that the $k_0^{(3)}$ contribution for this molecule is considerably larger than that of our ten-electron systems. We note that Pople *et al.* [115]) also observed that the $k_0^{(3)}$ contribution may be quite significant for some molecules. The fact that we have used a smaller basis for the BH calculation could influence the results to some extent (compare the basis set dependence of $H_2O$ in Table 5) but this does not seem to be so important. To clarify items (i) and (ii) it would be more convenient to compare our MB-RSPT calculations with full CI calculations, or at least with CI calculations truncated at quadruply excited configurations (CI-Q). Of the molecules studied in our paper[116]), we can make such a comparison for $H_2O$ in a double zeta basis[118]) and for BH (see Table 6). According to Hosteny *et al.* [118]), the CI-SD correlation contribution for $H_2O$ represents 94.9% in comparison with the CI-Q. For the BH molecule, CI-SD amounts to 95.4% of the correlation energy as compared with full CI[117]. Therefore, it seems to us that for the BH molecule, for which $k_0^{(3)}$ is significant, neither basis set dependence nor the $N^2$ behavior of CI-SD alone could explain the smaller amount of correlation energy included in $k_0^{(2)}$ and $k_0^{(3)}$ as compared to the CI-SD results. Instead, we think that the infinite summation of diagrams which include the effect of $T_2$ clusters in higher orders might be important.

To complete the description of the utility of the MB-RSPT, we would like to note that for the basis used, $k_0^{(2)}$ and $k_0^{(3)}$ calculations are relatively fast, *e.g.* for the largest basis set ($H_2O$) the integral transformation time is almost identical to that needed for the atomic integral calculation, while the time for the $k_0^{(2)}$ and $k_0^{(3)}$ calculation represents about 50% of what is needed for the SCF procedure.

## G. Method of Coupled-Pair Many-Electron Theory (CPMET) and Its Results

The system of Eqs. (137) and (138) is known as the CPMET. This method, as was demonstrated in Section V, considers only the doubly excited clusters. This assumption was only based on theoretical considerations. The full CI (FCI) calculation of Pipano and Shavitt[119] for $BH_3$ was a challenge to test CPMET numerically. Of course, a comparison with FCI results required an extension of CPMET to account for the $T_1$, $T_1T_2$ and $T_3$ clusters. This extended version of CPMET is called[51] the ECPMET. Testing of the ECPMET was performed[51] with the same basis set as the FCI calculation of Pipano and Shavitt. It was shown that the effect of linked four electron and higher clusters (both unlinked and linked) corresponds to 0.002% of the total correlation energy given by this basis set. The effect of $T_1$ clusters is very small (less than 0.1% of the total computed correlation energy). The approximation $T \approx T_2 \left( e^T = 1 + T_2 + \frac{1}{2} T_2^2 \right)$ gives a correlation energy almost identical to that obtained by the 0 + 2 + 4 excitation CI in which linked and unlinked contributions are lumped together. (The contribution of neglected linked clusters amounts to 0.004%). Also, the effect of the relative importance of linked ($T_3$) and unlinked ($T_1T_2$) clusters was tested. It was concluded that, as far as the calculation of the correlation energy is concerned, the relative importance of linked and unlinked terms for tri-excited clusters is just the opposite of that found for tetra-excited clusters: the role of unlinked tri-excited clusters is negligible compared to the linked tri-excited terms. Thus, the contribution of tri-excited clusters is predominantly due to linked terms. The overall contribution of tri-excited clusters was computed to be less than 0.8% of the total correlation energy.

As regards the computational aspects, in the CPMET we have to solve a system of Eqs. (137) and (138), which is a system of nonlinear equations of the form

$$\sum_j a_{ij} x_j + \sum_{j<k} b_{ijk} x_j x_k + c_i = 0 \tag{179}$$

where $x_i$ stands for $\mathbf{t}_2$-matrix elements, $c_i$ stands for v-matrix elements and $a_{ij}$ and $b_{ijk}$ are coefficients containing f- and v-matrix elements. The index $i$ refers to all distinct doubly excited configurations. This is a major advantage of CPMET because the order of a CPMET problem is considerably lower than the order of the corresponding $0 + 2 + 4$ CI problem. For example, in the case of $BH_3$, the order of the former is 21 whereas that of the latter is 88. The system of Eq. (179) can be solved iteratively. CPMET is obviously a nonvariational method. However, the advantage of having an upper bound to the energy is probably not as important when the method is accurate enough to give the correlation energy with an accuracy of a few percent.

## H. Variational Perturbation Calculations

In Section V we presented a method which is neither perturbational nor variational but which possesses certain features of both. Here, we shall make a few remarks on another promising method of this kind. Instead of the cluster expansion of the wave function this method is based on the idea of direct combination of the perturbation expansion with the variational method.

The Rayleigh-Schrödinger perturbation expansion for the exact wave function to the first order is given by

$$|\Psi_0\rangle = |\Phi_0\rangle + Q_0 W |\Phi_0\rangle \tag{180}$$

where $Q_0$, as was already shown, is given by

$$Q_0 = \frac{1 - |\Phi_0\rangle\langle\Phi_0|}{e_0 - H_0} = \frac{R_0}{e_0 - H_0} \tag{181}$$

Here W is the perturbation given by Eq. (67) and $e_0$ is defined by Eq. (63). The operator $Q_0$ generates the excited configurations and, by its effect, the doubly excited configurations enter the expansion (180). As we already know, the doubly excited configurations give a dominant contribution to the correlation energy. However, any CI wave function restricted to (singly and) doubly excited configurations exhibits a few unpleasant properties. Among these is the failure to approximate the exact ground state energy $E_0$ for larger internuclear distances in dissociation processes (see Section VI.I.). The function (180), of course, possesses the same shortcoming. To avoid this difficulty, Cederbaum *et al.* [120] modified its form somewhat

$$|\Psi_0\rangle = |\Phi_0\rangle + \gamma \frac{R_0}{\epsilon - H_0} W |\Phi_0\rangle \tag{182}$$

by introducing new variational parameters $\gamma$ and $\epsilon$. The idea used here is similar to that used by Goldhammer and Feenberg[121] for the refinement of the Brillouin-Wigner perturbation expansion. The function (182) is used to determine the ground state energy

$$E_0 = \frac{\langle\Psi_0|H|\Psi_0\rangle}{\langle\Psi_0|\Psi_0\rangle} \tag{183}$$

which leads to the following equation

$$E_0 = e_0 + \langle \Phi_0 | \mathrm{W} | \Phi_0 \rangle + \frac{1}{2}\left( \frac{E_3 - E_2}{S} + \epsilon - e_0 \right)$$
$$- \left[ \frac{1}{4}\left( \frac{E_3 - E_2}{S} + \epsilon - e_0 \right)^2 + \frac{E_2^2}{S} \right]^{1/2} \qquad (184)$$

where

$$E_2 = \left\langle \Phi_0 | \mathrm{W} \frac{\mathrm{R}_0}{\epsilon - \mathrm{H}_0} \mathrm{W} | \Phi_0 \right\rangle \qquad (185)$$

$$S = \left\langle \Phi_0 | \mathrm{W} \frac{\mathrm{R}_0}{(\epsilon - \mathrm{H}_0)^2} \mathrm{W} | \Phi_0 \right\rangle \qquad (186)$$

$$E_3 = \left\langle \Phi_0 | \mathrm{W} \frac{\mathrm{R}_0}{\epsilon - \mathrm{H}_0} (\mathrm{W} - \langle \Phi_0 | \mathrm{W} | \Phi_0 \rangle) \frac{\mathrm{R}_0}{\epsilon - \mathrm{H}_0} \mathrm{W} | \Phi_0 \right\rangle \qquad (187)$$

Equation (184) was used[120)] to calculate the correlation energy of the $H_2O$ molecule. Compared to the CI calculation of Diercksen and Sutcliffe[122)], which included all singly and doubly excited configurations (2063 configurations), it gave 97.3% of the correlation energy for exactly the same basis set and only one variational parameter ($\epsilon$ was fixed at $e_0$). Equation (184) was also applied[120)] to the potential curve of $N_2$. The calculated correlation energy was demonstrated to have reasonable behavior even at larger internuclear distances, where the RS expansion up to third order breaks down. The same result was obtained[123)] with the model calculations for the anharmonic oscillator.

Cederbaum and Schönhammer generalized[124)] the variational trial function (182) by replacing the perturbation expansion of the exact wave function to the first order [Eq. (182)] with a constructed variational "ansatz" for a wave function expanded through all orders of the perturbation theory. As they demonstrated, this approach includes, as special cases, the Goldhammer-Feenberg[121)] refinement of the Brillouin-Wigner series, the analogous refinement of the Rayleigh-Schrödinger series and the continued-fraction expansion of the resolvent matrix element[124)].

The reason that we have paid attention to this method is the following: Performing the perturbation calculations, one very often faces the situation that the perturbation is not small. In these cases the perturbation theory can only be used if one goes to higher orders of perturbation expansion or if partial summations of an infinite series of some preselected contributions are performed. To go to higher orders of perturbation expansion is very unpleasant from the computational point of view. To perform partial infinite summations of some preselected contributions can also be sometimes complicated by, among other things, the question of which contributions to sum to infinity. The variational-perturbational method, just mentioned, seems to be very promising in this respect because instead of performing the infinite summations of some preselected contributions, one could perform variational-perturbational calculations which are much simpler.

## I. Potential Curves and Correlation Energy

Let us start the discussion with Fig. 15, which presents various levels of the approach to the potential curve of $F_2$. The Hartree-Fock approximation is especially poor in this case because it does not lead to a correct dissociation limit, *i.e.* to the atoms in the Hartree-Fock ground states. In the case of $F_2$, the proper dissociation is achieved

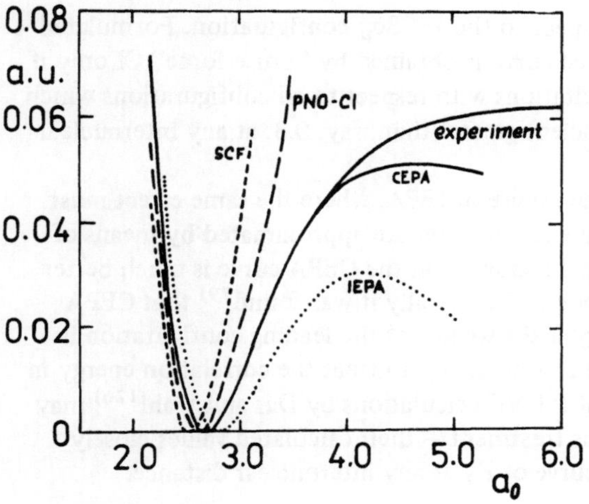


Fig. 15. Potential curves of $F_2$ with different approximations[79)]

by adding the excited configuration $1\pi^4 3\sigma_u^2$ to the ground state $1\pi^4 3\sigma_g^2$. Generally, however, it is not sufficient to include just one configuration in the ground state wave function. For example, to ensure a correct dissociation, it is necessary to consider[68)] 3 excited configurations for the $C_2 \longrightarrow 2\,C(^3P)$ process and 9 excited configurations for the $N_2 \longrightarrow 2\,N(^4S)$ process. This type of conventional CI wave function was referred to by Lie and Clementi[68)] as the "Hartree-Fock with proper dissociation" (HFPD). Although the HFPD functions give physically reasonable curves in the sense that at dissociation limits the energy is equal to the sum of the ground state SCF energies of the respective atoms, they still give poor dissociation energies. As seen in Fig. 15, the PNO-CI curves do not provide a reasonable dissociation energy, although the PNO-CI is considerably superior to the simple HFPD function. The use of the pair natural orbitals is not essential here and for the purpose of the present discussion the PNO-CI function may be considered as a conventional CI function restricted to doubly excited configurations. One of the failures of this function may be understood after inspection of Table 7. It can be seen that as the internuclear distance is increased, the coefficient of the $1\pi^4 3\sigma_u^2$ configuration increases and the function of the ground state $1\pi^4 3\sigma_g^2$ loses its role as the leading configuration. As a consequence, for a correct evaluation of the correlation energy it is no longer sufficient to assume double excitations only with respect to the configuration $1\pi^4 3\sigma_g^2$ but one must also take into consideration the configuration $1\pi^4 3\sigma_u^2$. In the language of the traditional CI it is therefore necessary to also assume quadruply excited states

Table 7. Coefficients ($C^2$) of $1\pi^4 3\sigma_g^2$ and $1\pi^4 3\sigma_u^2$ configurations in the PNO-CI wave function[79] of $F_2$

| $r$ (in $a_0$) | 2.6 | 3.0 | 3.5 | 4.0 |
|---|---|---|---|---|
| $(1\pi^4 3\sigma_g^2)$ | 0.920 | 0.900 | 0.844 | 0.765 |
| $(1\pi^4 3\sigma_u^2)$ | 0.1539 | 0.2268 | 0.3333 | 0.43654 |

that are double excitations with respect to the $1\pi^4 3\sigma_u^2$ configuration. Formulated generally[125], "a reasonable potential curve is obtained by "brute force" CI only if it includes all single and double excitations with respect to all configurations which enter the wave function with coefficient greater than, say, 0.3, at any internuclear separation".

We can now also understand the failure of IEPA, where the same effect must take place. In CEPA, the quadruply excited states are approximated by means of unlinked clusters. Accordingly, Fig. 15 shows that the CEPA curve is much better than the results of the other treatments. Empirically it was found[79] that CEPA overestimates the correlation energy if the weight of the leading configuration is decreased below 0.8. In such a case it is necessary to treat the correlation energy in a more rigorous manner. The MC SCF OVC calculations by Das and Wahl[126] may serve as an example of such rigorous treatment – their calculated values closely match the experimental potential curve of $F_2$ at any internuclear distance.

## VII. Use of the Many-Body Rayleigh-Schrödinger Perturbation Theory for Ionized and Excited States

In this chapter we present the utility of the MB-RSPT for applications in different fields of spectroscopy. The theory will be demonstrated to be an excellent tool for interpreting various phenomena related to ionization, excitation and combinations of both.

### A. Ionization Potentials

The simplest process, among those noted above, is ionization of a single electron. The true vertical ionization potential ($I_H$) for the ionization of an electron from the spin-orbital $|H\rangle$ is given by

$$I_H = -\epsilon_h - R + C \tag{188}$$

where $\epsilon_h$ is the orbital energy, $R$ stands for reorganization and $C$ for correlation effects. In molecules for which $-R + C$ effects almost cancel each other out, Koopmans' theorem ($I_H = -\epsilon_h$) gives a satisfactory account of observed ionization potentials. In cases where Koopmans' theorem does not hold, a more sophisticated approach must be used. Such approaches have already been attempted. Cederbaum *et al.*[127,128] did pioneering work using the Green's function approach.

Here we shall demonstrate how to obtain the explicit expression[129,130] for (188) by means of the MB-RSPT. Let us study the ionized state which we shall describe using $|\Psi_i\rangle$. We shall limit ourselves to a state $|\Psi_i\rangle$ for which the initial state $|\Phi_i\rangle$ is obtained from the neutral closed shell ground state in which we annihilate one particle. The state $|\Phi_i\rangle$ is therefore realized by

$$|\Phi_i\rangle = X_H|\Phi_0\rangle \tag{189}$$

The corresponding eigenvalue problems are

$$K|\Psi_i\rangle = k_i|\Psi_i\rangle \tag{190}$$

$$K_0|\Phi_i\rangle = \kappa_i|\Phi_i\rangle \tag{191}$$

where, for $k_i$, we have the expansion (52)

$$k_i = \kappa_i + \sum_{n=0}^{\infty} \langle\Phi_i|W[Q_i(W + \kappa_i - k_i)]^n|\Phi_i\rangle \tag{192}$$

The ionization potential $I_H$ is given by

$$I_H = k_i - k_0 \tag{193}$$

where $k_i$ is defined according to Eq. (190), and $k_0$ refers to the ground state. Up to third order we can write [using Eq. (189)]

$$\begin{aligned} I_H = {} & \{\langle\Phi_0|X_H^+K_0X_H|\Phi_0\rangle + \langle\Phi_0|X_H^+WQ_iWX_H|\Phi_0\rangle \\ & + \langle\Phi_0|X_H^+WQ_iWQ_iWX_H|\Phi_0\rangle\} - \{\langle\Phi_0|WQ_0W|\Phi_0\rangle \\ & + \langle\Phi_0|WQ_0WQ_0W|\Phi_0\rangle\} \end{aligned} \tag{194}$$

The first term of Eq. (194) gives us $-\epsilon_h$ and corresponds therefore to Koopmans' theorem. Let us write Eq. (194) formally as

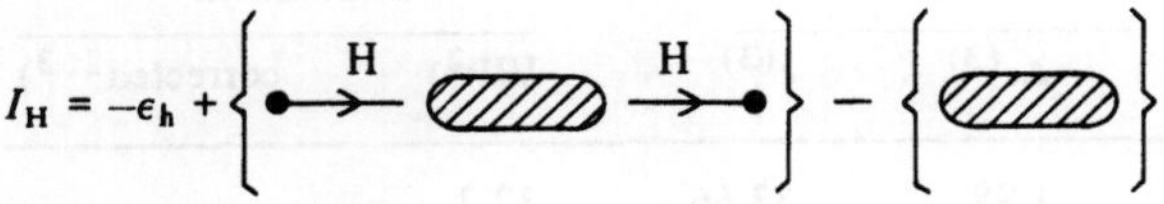

(195)

where 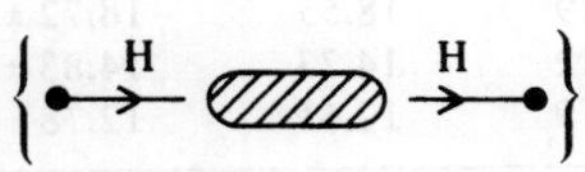 stands for the second and the third term of (194),

and  stands for last two terms of (194). In order to calculate (194), we know that all the operators in individual terms have to be contracted. With respect to the terms containing $X_H$ and $X_H^+$ operators, the simplest among possible contractions is the following one

$$\langle\Phi_0|X_H^+ W Q_i W X_H|\Phi_0\rangle + \langle\Phi_0|X_H^+ W Q_i W Q_i W X_H|\Phi_0\rangle \quad (196)$$

We note that contractions coupling the $X_H^+$ and $X_H$ operators gives us Kronecker delta $\delta_{HH}$. By coupling W operators in both terms we refer to all possible contractions. Schematically, 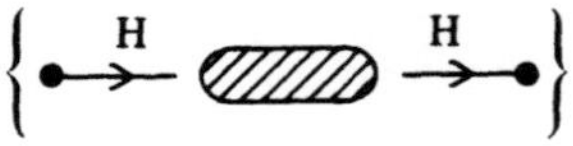 can be decomposed in the following way:

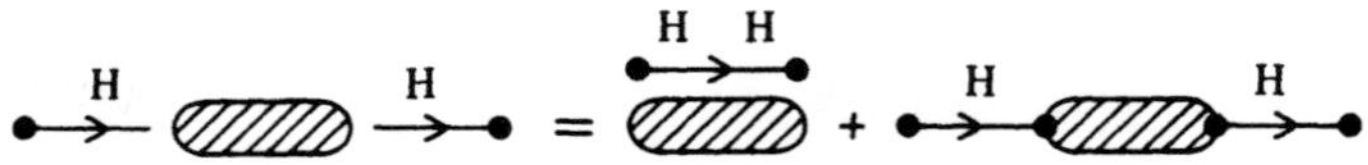 (197)

The first term of (197) stands for contraction (196) and the second term represents diagrams where operators $X_H^+$ and $X_H$ are directly contracted with W operators. Substituting (197) into (195) gives us

$I_H = -\epsilon_h +$  (198)

because the first term of (197) cancels out with the ground state correlation energy diagrams (it differs from 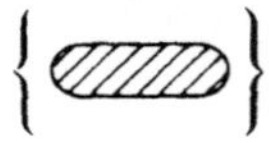 only by a multiplicative factor $\delta_{HH}$). The corresponding second and third order Hugenholtz diagrams are presented in Ref.[130]. To illustrate the utility of the method we present the results for the $H_2O$ molecule in Table 8.

Table 8. Vertical ionization potentials (in eV) for $H_2O$[130]

| Ionization | Koopmans | Calculated[1] | | | Experiment | |
|---|---|---|---|---|---|---|
| | $-\epsilon$ | $\Delta\epsilon^{(2)}$ | $\Delta\epsilon^{(3)}$ | $\epsilon^{(3)}$ | VIP[2] | corrected[2,3] |
| $2a_1$ | 36.56 | −4.87 | 1.98 | 33.66 | 32.2 | |
| $1b_2$ | 19.30 | −1.49 | 1.39 | 19.19 | 18.55 | 18.72 ± 0.22 |
| $3a_1$ | 15.66 | −2.46 | 1.91 | 15.12 | 14.73 | 14.83 ± 0.11 |
| $1b_1$ | 13.66 | −2.91 | 2.05 | 12.79 | 12.61 | 12.78 |

[1]) Values up to the third order, $\epsilon^{(3)}$, and second and third order contributions, $\Delta\epsilon^{(2)}$ and $\Delta\epsilon^{(3)}$.
[2]) Vertical ionization potentials.
[3]) Corrected for vibrational effects.

### B. Electron Affinities

The procedure described for ionization potentials can serve us as a flow-chart to obtain other quantities $k_i$ by realizing $|\Phi_i\rangle$ in different ways. For electron affinities instead of (189) we have

$$|\Phi_i\rangle = X_P^+ |\Phi_0\rangle \tag{199}$$

The electron affinity is given by

$$A_P = k_i - k_0 \tag{200}$$

Using the same arguments as for Eq. (193), we arrive at the expression:

$A_P = \epsilon_P +$  (201)

where the corresponding Hugenholtz diagrams for the second term of Eq. (201) are given in Ref.[129]. Generally, the diagrams for electron affinities may be inferred from diagrams of ionization potentials[130] by changing the hole lines $H$ to particle lines $P$.

### C. Excitation Energies

Explicit expressions for excitation energies were derived[37] by making use of the Green's function approach. As was demonstrated[9, 37], the Green's function approach yields the same expressions for the excitation energies as time–independent MB-RSPT. The explicit formulas up to the third order of the MB-RSPT expansion for excitation energies are presented in Refs.[9, 37]. As we shall demonstrate here, the ionization potentials, as well as, the electron affinities are implicitly included in excitation energies. This can also be demonstrated diagrammatically because our diagrams[130] for ionization potentials represent the part of the self-energy diagrams for the two-particle Green's function[37)f)]. Let us study the system which we obtain from a neutral closed shell ground state by exciting one electron from the occupied spin-orbital $|H\rangle$ to the unoccupied spin-orbital $|P\rangle$. We shall describe this state using $|\Psi_i\rangle$. The state $|\Phi_i\rangle$ is now realized by

$$|\Phi_i\rangle = X_P^+ X_H |\Phi_0\rangle \tag{202}$$

---

f) To be more specific, all our third order diagrams for ionization potentials[130] are simply time-versions of the Diagrams 12, 13 and 14 in Ref.[37]. As we have already said Cederbaum *et al.*[127, 128] have used the one-particle Green's function technique to obtain the explicit expression for ionization potentials. From what has been said above, it is understandable that there is a one to one correspondence between our diagrams[130] and those labeled as A1-A6, C1-C6 and D1-D6 in Ref.[131]. Cederbaum's approach [127, 128, 131] is the Green's function technique in Brillouin-Wigner perturbative form.

We assume $|\Phi_i\rangle \longrightarrow |\Psi_i\rangle$ if the perturbation is switched on. In analogy to (193), the excitation energy $E_{H\to P}$ is given by

$$E_{H\to P} = k_i - k_0 \tag{203}$$

Up to the third order we can write

$$\begin{aligned} E_{H\to P} = {} & \langle\Phi_0|X_H^+ X_P K_0 X_P^+ X_H|\Phi_0\rangle + \langle\Phi_0|X_H^+ X_P W X_P^+ X_H|\Phi_0\rangle \\ & + \langle\Phi_0|X_H^+ X_P W Q_i W X_P^+ X_H|\Phi_0\rangle + \langle\Phi_0|X_H^+ X_P W Q_i (W \\ & - \langle\Phi_0|X_H^+ X_P W X_P^+ X_H|\Phi_0\rangle) Q_i W X_P^+ X_H|\Phi_0\rangle \\ & - \langle\Phi_0|W Q_0 W|\Phi_0\rangle - \langle\Phi_0|W Q_0 W Q_0 W|\Phi_0\rangle \end{aligned} \tag{204}$$

The first term of Eq. (204) gives us $-\epsilon_h + \epsilon_p$. Formally, we shall write Eq. (204) in the following way:

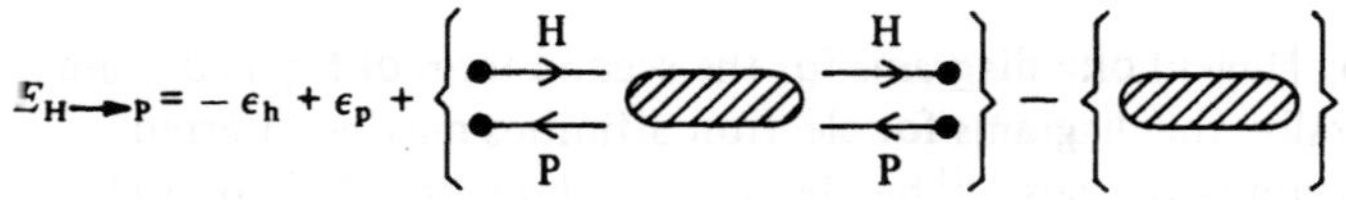


(205)

The last term of Eq. (205) stands for the last two terms of Eq. (204) and the third term of (205) stands for the rest terms of Eq. (204). In analogy to the treatment of ionization potentials we can carry out the following decomposition

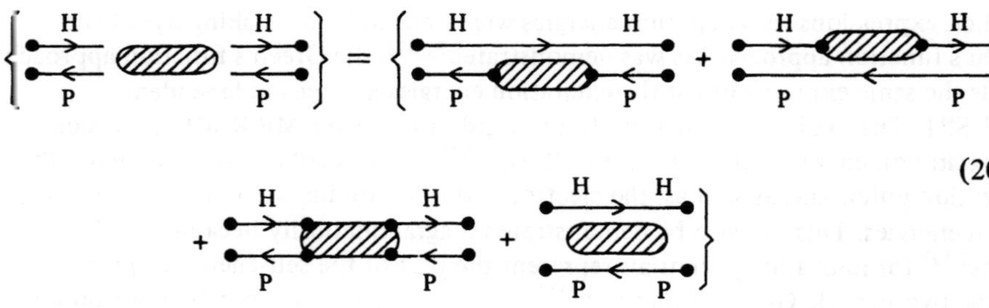


(206)

On substituting (206) into (205) we can write

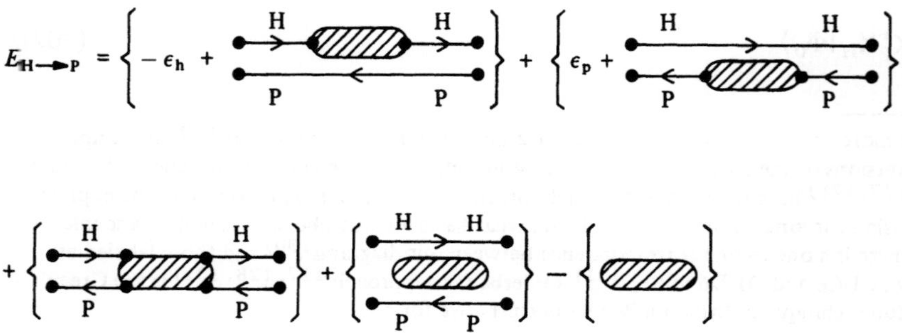


(207)

The last two terms of Eq. (207) cancel each other out once again and by comparing (198) and (201) we get the final form

$E_{H \rightarrow P} = I_H + A_p +$ 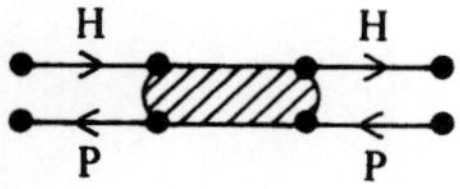 (208)

where the last term represents diagrams in which both $X_H^+ X_P$ and $X_H X_P^+$ are coupled with the W operators in Eq. (204). Equation (208) documents the fact that the ionization potentials and electron affinities are included in excitation energies, as was mentioned above.

The state realized by Eq. (208) does not represent any real spectroscopic state. To arrive at observable spectral states we have to substitute the corresponding eigenstates of $S^2$ and $S_z$ operators for $|\Phi_i\rangle$. We note that the calculation of excitation energies can be slightly more complicated when the unperturbed functions $|\Phi_i\rangle$ are degenerate. In that case we have to use the degenerate MB-RSPT[9, 41, 132–134]. The degenerate MB-RSPT has successfully been applied to $H_2$ and BH molecules[132–134] (the calculated excitation energies were within 0.1 eV).

## D. Double Ionizations and Shake-up Processes

We have already shown that excitation energies can be diagrammatically decomposed to yield simpler quantities such as ionization potentials and electron affinities plus some remaining diagrams. MB-RSPT permits the use of this treatment for even more complex processes. In this section, we present the applicability of the theory to double ionizations observed in Auger spectra as well as excitations accompanying photoionization (shake-up processes) observed in ESCA and photoelectron spectroscopy. A detailed description of this approach is given in Refs.[135, 136]. Here we shall present only the formal description.

For a double ionization we have:

$$|\Phi_i\rangle = X_I X_H |\Phi_0\rangle \qquad (209)$$

Using arguments similar to those regarding Eq. (203) for the excitation energies, we obtain the double ionization energy

$E_{I,H} = \left\{ -\epsilon_i +\right.$ 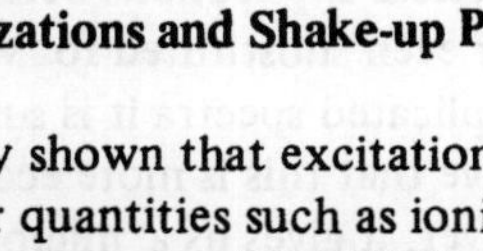 $\left.\right\} +$ 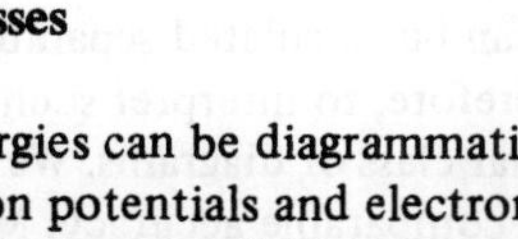 (210)

$+ \left\{ -\epsilon_h +\right.$ 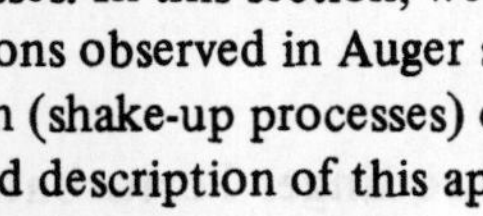 $\left.\right\} +$ 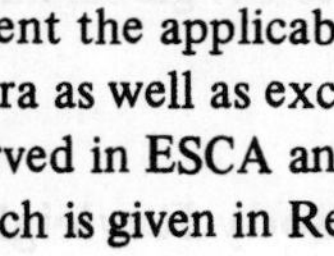 $-$ 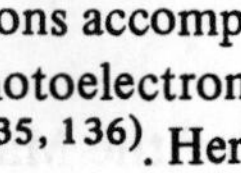

which may be reduced to

$E_{I,H} = I_I + I_H +$ 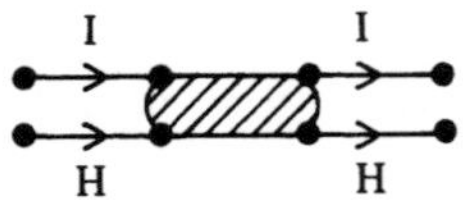 (211)



In the case of shake-up processes, the state $|\Phi_i\rangle$ can be realized as

$$|\Phi_i\rangle = X_K^+ X_I X_H |\Phi_0\rangle \tag{212}$$

The corresponding energy is given by

$E_{H,I\rightarrow K} = +\epsilon_k - \epsilon_i - \epsilon_h +$ 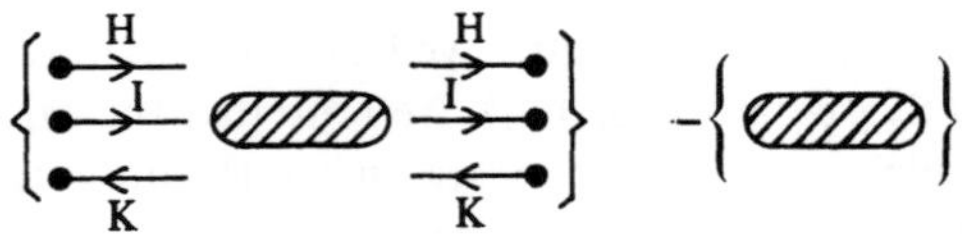 (213)



The first term in the curled brackets can be decomposed to quantities which have already been presented, *i.e.* ionization potentials, electron affinities, excitation energies, double ionization energies as well as certain remaining terms. This composite property of MB-RSPT might be of value for practical applications because the mentioned quantities can be calculated separately or even substituted for with experimental values. Therefore, to interpret such complicated spectra it is sufficient to calculate a particular class of diagrams. We believe that this is more economic than a CI calculation of comparable accuracy. Moreover, it gives us a microscopic view of the problem in contrast to the global nature of the CI calculation.

## VIII. Use of the Many-Body Rayleigh-Schrödinger Perturbation Theory for the Interaction Between Two Closed-Shell Systems

In this chapter we shall present the fundamental ideas of the MB-RSPT in its application to the interaction of two closed-shell ground state molecular systems. Remarkable features of the approach are the following:

(i) the antisymmetry property of unperturbed wave function for supersystem $A \cdots B$ can be treated properly;

(ii) the nonorthogonality problem of the basis set for supersystem can be treated properly;

(iii) simple physical interpretation of terms appearing in explicit formula for the interaction energy.

Owing to the complicated nature of the problem, we are forced to change, to some extent, the notation for the spin-orbitals due to reasons that will be understandable from the following remarks.

Let us study the interaction of two closed shell molecular systems $A$ and $B$ in the ground states. First, we summarize the main ideas which we made use of in our derivation[139]. The interaction energy between systems $A$ and $B$ is defined as the exact energy of the Hamiltonian H of the supersystem $A \cdots B$ from which we substract the exact energies of the Hamiltonians $\mathsf{H}_A$ and $\mathsf{H}_B$ (*i.e.* exact energies of the systems $A$ and $B$). In Chapter IV we have demonstrated how to calculate the exact energies through MB-RSPT expansion. This expansion was given in terms of creation and annihilation operators which were defined on the orthonormal basis set. If MB-RSPT is to be applied to the supersystem $A \cdots B$, we encounter the problem that the basis set constituted from eigenfunctions of systems $A$ and $B$ is generally not orthogonal. Therefore, this basis has to be orthogonalized so that the Hamiltonian can be written in the second quantization formalism. We choose an orthogonalization which permits distinguishing the $\mathsf{H}_A$ and $\mathsf{H}_B$ defined over their own basis sets in the Hamiltonian of the supersystem. Then, choosing the sum of the Hartree-Fock orbital energies for systems $A$ and $B$ as $H_0$ (unperturbed Hamiltonian) for the supersystem $A \cdots B$ and defining the rest as perturbation, we can apply the MB-RSPT as we did in Chapter IV.

We label the spin-orbitals of system $A$ as

$$\chi_A \equiv \{|\varphi_i\rangle; i \in A\} \tag{214}$$

and those of system $B$ as

$$\chi_B \equiv \{|\varphi_j\rangle; j \in B\} \tag{215}$$

The matrix of the overlap integrals between the systems $A$ and $B$ (for a particular distance $R$) is defined by

$$\mathbf{S}_{AB} = \chi_A^+ \chi_B \equiv \{S_{ij} = \langle\varphi_i|\varphi_j\rangle; i \in A, j \in B\} \tag{216}$$

Using the column vectors (214) and (215) we construct the new column vector

$$\chi = \begin{pmatrix} \chi_A \\ \chi_B \end{pmatrix} \tag{217}$$

Let us assume that there exists a non-singular matrix **A** which transforms basis $\chi$ into orthonormal basis set $\gamma$,

$$\gamma = \mathbf{A}\chi \equiv \{|\gamma_i\rangle; i \in A + B\} \tag{218}$$

and which also satisfies the following asymptotic condition

$$\lim_{R \to \infty} \mathbf{A} = \mathbf{I} \tag{219}$$

The spin-orbitals $\{|\gamma_i\rangle; i \in A + B\}$ form a basis set for the supersystem $A \cdots B$. One possible procedure of realizing the transformation (218) as well as (219) is the Löwdin symmetric orthonormalization method[137)].

We can write

$$\mathbf{A} = (\mathbf{I} + \mathbf{S})^{-1/2} \tag{220}$$

where

$$\mathbf{S} = \begin{pmatrix} 0 & \mathbf{S}_{AB} \\ \mathbf{S}_{BA} & 0 \end{pmatrix} \tag{221}$$

Assuming that all the absolute values of the eigenvalues of matrix **S** are less than one[137], we can express the right hand side of Eq. (220) in a power series of **S**

$$\mathbf{A} = \sum_{n=0}^{\infty} \binom{n}{-1/2} \mathbf{S}^n = 1 - \frac{1}{2}\mathbf{S} + \frac{3}{8}\mathbf{S}^2 - \ldots \tag{222}$$

From (218) we obtain an expression for the individual components of the column vector $\gamma$

$$|\gamma_i\rangle = |\varphi_i\rangle - \frac{1}{2} \sum_{j \in B} S_{ij} |\varphi_j\rangle + \frac{3}{8} \sum_{k \in B} \sum_{l \in A} S_{ik} S_{kl} |\varphi_l\rangle \tag{223}$$

where $i \in A$. For $i \in B$ one has to exchange $A \longleftrightarrow B$ in Eq. (223). Equation (223) has a very important property which permits us to "see" the Hamiltonians $\mathsf{H}_A$ and $\mathsf{H}_B$ in the Hamiltonian for the supersystem $\mathsf{H}$.

According to Eq. (29), for the Hamiltonian of the supersystem we have

$$\mathsf{H} = \sum_{i,j \in A+B} \langle \gamma_i | \mathsf{z} | \gamma_j \rangle \mathsf{X}_i^+ \mathsf{X}_j + \frac{1}{2} \sum_{ijkl \in A+B} \langle \gamma_i \gamma_j | \mathsf{v} | \gamma_k \gamma_l \rangle \mathsf{X}_i^+ \mathsf{X}_j^+ \mathsf{X}_l \mathsf{X}_k \tag{224}$$

We can now substitute for $|\gamma_i\rangle$ from Eq. (223). This brings about the split

$$\langle \gamma_i | \mathsf{z} | \gamma_j \rangle = \langle \varphi_i | \mathsf{z} | \varphi_j \rangle + \Delta \langle \varphi_i | \mathsf{z} | \varphi_j \rangle \tag{225}$$

and similarly

$$\langle \gamma_i \gamma_j | \mathsf{v} | \gamma_k \gamma_l \rangle = \langle \varphi_i \varphi_j | \mathsf{v} | \varphi_k \varphi_l \rangle + \Delta \langle \varphi_i \varphi_j | \mathsf{v} | \varphi_k \varphi_l \rangle \tag{226}$$

where $\Delta$ terms are due to the second and the remaining terms of the expansion (223). The first terms in Eqs. (225) and (226) are terms which permit us to isolate the systems $A$ and $B$ over $|\varphi_i\rangle$ basis sets and thus to formulate simply the unperturbed Hamiltonian and the perturbation. For further considerations it is appropriate to introduce an unperturbed ground-state vector $|\Phi_0^{A \cdots B}\rangle$ for the supersystem $A \cdots B$

$$|\Phi_0^{A \cdots B}\rangle = \prod_{i \in A}^{\text{occ}} \mathsf{X}_i^+ \prod_{j \in B}^{\text{occ}} \mathsf{X}_j^+ |0\rangle \tag{227}$$

In order to apply the MB-RSPT we have to change our Hamiltonian (224) to take the form

$$\mathsf{H} = \mathscr{E}_0 + \mathsf{H}_0 + \mathsf{W} \tag{228}$$

where $\mathscr{E}_0$ is a scalar quantity, $\mathsf{H}_0$ is the unperturbed Hamiltonian for the supersystem and $\mathsf{W}$ is the perturbation. We can arrive at this form of $\mathsf{H}$ by applying Wick's theorem to Eq. (224) and by satisfying the relations (225) and (226). For the scalar quantity $\mathscr{E}_0$ we obtain

$$\mathscr{E}_0 = E_A^{\mathrm{HF}} + E_B^{\mathrm{HF}} + e_0 \tag{229}$$

where $E_A^{\mathrm{HF}}$ and $E_B^{\mathrm{HF}}$ respectively, are the Hartree-Fock energies of the systems $A$ and $B$ and $e_0$ comprises scalar quantities that remain after extraction of $\mathsf{H}_A$ and $\mathsf{H}_B$ from $\mathsf{H}$ (Eq. (224)). We choose $\mathsf{H}_0$ to be the sum of the Hartree-Fock orbital energies of systems $A$ and $B$,

$$\mathsf{H}_0 = \sum_{i \in A} \epsilon_i N[\mathsf{X}_i^+ \mathsf{X}_i] + \sum_{j \in B} \epsilon_j N[\mathsf{X}_j^+ \mathsf{X}_j] \tag{230}$$

The other remaining terms are taken as perturbation. Now we can apply the MB-RSPT expansion from Chapter IV.

Thus, we obtain the energy of the supersystem, $E^{AB}$, from which it is possible to extract exact energies $E_A$ and $E_B$ for systems $A$ and $B$ and to obtain the expression for the interaction energy up to a particular order of the perturbation expansion. It was not the aim of this chapter to present the details of the MB-RSPT treatment of intermolecular interactions but rather to point out the fundamental ideas involved. Detailed derivation and explicit formulas as well as the physical interpretation of individual terms are presented in Refs.[138, 139].

*Acknowledgements.* The stimulus for this paper arose from suggestions by Dr. Zahradník. His interest, encouragement and valuable comments on this manuscript are sincerely acknowledged. For technical assistance we also wish to express our appreciation to Mrs. Žohová and Mrs. Týleová. One of the authors (I.H.) wishes to express his sincere gratitude to members of the Quantum Theory Group, Department of Applied Mathematics, Faculty of Mathematics, University of Waterloo, Ontario, Canada, who familiarized him during his stay at QTG in 1969–1971 with the theoretical apparatus used in this review.

Note Added in Proof

We wish to extend our acknowledgements to Professor Ahlrichs for reading the preprint. In his opinion our discussion on pages 44–46 is too conclusive with respect to the justifiability of CEPA. He suggests that the arguments on p. 46 should be so reformulated that it is possible to assume such an unitary transformation of occupied and virtual orbitals which would conform best to the approximations inherent in Eqs. (170) and (172).

## IX. References

1) Klessinger, M.: Fort. Chem. Forsch. *9*, 354 (1968)
2) Kutzelnigg, W., Del Re, G., Berthier, G.: Topics Curr. Chem. *22*, 1 (1971)
3) Kutzelnigg, W.: Topics Curr. Chem. *41*, 31 (1973)
4) Parr, R. G.: Quantum theory of molecular electronic structure. New York: Benjamin 1963
5) Schaefer III, H. F.: The electronic structure of atoms and molecules. A survey of rigorous quantum mechanical results. Reading, Massachusetts: Addison-Wesley 1972

6) V. Herigonte, P.: Struct. and Bonding *12,* 1 (1972)
7) Van der Velde, G. A.: Dissertation, Groningen 1974
8) Kutzelnigg, W.: Methoden zur Berücksichtigung der Elektronenkorrelation. Bochum: Ruhr-Universität 1976
9) Paldus, J., Čížek, J.: Advan. Quant. Chem. *9,* 105 (1975)
10) Shavitt, I., in: Modern theoretical chemistry. Schaefer, H. F. (ed.). New York: Plenum Press 1976
11) Davidson, E. R.: Configuration interaction description of electron correlation. In: The world of quantum chemistry. Proceedings of the First International Congress on Quantum Chemistry. Daudel, R., Pullman, B. (eds.). Dordrecht (Holland): D. Reidel *1974,* pp. 17–30
12) Bagus, P. S., Liu, B., McLean, A. D., Yoshimine, M.: *Ab initio* computation of molecular structures through configuration interaction. In: Computational methods for large molecules and localized states in solids. Herman, F., McLean, A. D., Nesbet, R. K. (eds.). New York: Plenum Press 1973, pp. 87–115
13) Buenker, R. J., Peyerimhoff, S. D.: Theoret. Chim. Acta *35,* 33 (1974)
14) Nesbet, R. K.: J. Chem. Phys. *36,* 1518 (1962)
15) McLean, A. D.: J. Chem. Phys. *39,* 2653 (1963)
16) Nesbet, R. K.: Advan. Chem. Phys. *9,* 321 (1965)
17) Snyder, L. C.: J. Chem. Phys. *46,* 3602 (1967)
18) Snyder, L. C., Basch, H.: J. Amer. Chem. Soc. *91,* 2189 (1969)
19) Ahlrichs, R.: Theoret. Chim. Acta *35,* 59 (1974)
20) Chapter V in Ref.[5]
21) Clementi, E., Gayles, J. N.: J. Chem. Phys. *47,* 3837 (1967)
22) Goldfinger, P., Verhaegen, G.: J. Chem. Phys. *50,* 1467 (1969)
23) Hehre, W. J., Ditchfield, D., Radom, L., Pople, J. A.: J. Amer. Chem. Soc. *92,* 4796 (1970)
24) Hehre, W. J., McIver, R. T., Jr., Pople, J. A., Schleyer, P. v. R.: J. Amer. Chem. Soc. *96,* 7162 (1974)
25) Radom, L.: J. Chem. Soc., Chem. Comm. *1974,* 403
26) Matcha, R. L.: J. Chem. Phys. *48,* 335 (1968)
27) Radom, L., Hehre, W. J., Pople, J. A.: J. Chem. Soc., Inorg. Phys. Theor. *1971,* 2299
28) Radom, L., Hehre, W. J., Pople, J. A.: J. Amer. Chem. Soc. *93,* 289 (1971)
29) Wahl, A. C.: J. Chem. Phys. *41,* 2600 (1964)
30) Hurley, A. C.: Advan. Quant. Chem. *7,* 315 (1973)
31) Hariharan, P. C., Pople, J. A.: Theoret. Chim. Acta *28,* 213 (1973)
32) Ahlrichs, R., Driessler, F., Lischka, H., Staemmler, V., Kutzelnigg, W.: J. Chem. Phys. *62,* 1235 (1975)
33) March, N. H., Young, W. H., Sampanthar, S.: The many-body problem in quantum mechanics. Cambridge: University Press 1967
34) Raimes, S.: Many-electron theory. Amsterdam-London: North-Holland Publishing Company 1972
35) Roman, P.: Advanced quantum theory. Reading, Massachusetts: Addison-Wesley 1965
36) Schweber, S. S.: An introduction to relativistic quantum field theory. Evanston, Illinois: Row, Peterson and Co. 1961
37) Paldus, J., Čížek, J.: J. Chem. Phys. *60,* 149 (1974)
38) Paldus, J., Wong, H. C.: Comp. Phys. Comm. *6,* 1 (1973)
39) Wong, H. C., Paldus, J.: Comp. Phys. Comm. *6,* 9 (1973)
40) Kaldor, U.: J. Comp. Phys. *20,* 432 (1976)
41) Brandow, B. H.: Rev. Mod. Phys. *39,* 771 (1967)
42) Fetter, A. L., Walecka, J. D.: Quantum theory of many-particle systems. New York: McGraw-Hill 1971
43) Gell-Mann, M., Brueckner, K.: Phys. Rev. *106,* 364 (1957)
44) Kelly, H. P.: Phys. Rev. *B136,* 896 (1964)
45) Kelly, H. P.: Phys. Rev. *144,* 39 (1966)
46) Claverie, P., Diner, S., Malrieu, J. P.: Intern. J. Quantum Chem. *1,* 751 (1967)
47) Sinanoğlu, O.: J. Chem. Phys. *36,* 706 (1962)
48) Čížek, J.: J. Chem. Phys. *45,* 4256 (1966)

49) Primas, H.: Separability in many-electron systems. In: Modern quantum chemistry. Istanbul lectures. Sinanoğlu, O. (ed.). New York: Academic Press 1965
50) Paldus, J., Čížek, J.: Relation of coupled-pair theory, CI and some other many-body approaches. In: Energy, Structure, and Reactivity. Proceedings of the 1972 Boulder Summer Research Conference on Theoretical Chemistry. Smith, D. W., McRae, W. B. (eds.). New York: Wiley 1973, pp. 198–212
51) Paldus, J., Čížek, J., Shavitt, I.: Phys. Rev. *A5*, 50 (1972)
52) Čížek, J., Paldus, J.: Intern. J. Quantum Chem. *5*, 359 (1971)
53) Hylleraas, E. A.: Z. Phys. *54*, 347 (1929)
54) Hurley, A. C., Lennard-Jones, J. E., Pople, J. A.: Proc. Roy. Soc. (London) *A220*, 446 (1953)
55) Kapuy, E.: Acta Phys. Acad. Sci. Hung. *9*, 237 (1958)
56) Parks, J. M., Parr, R. G.: J. Chem. Phys. *28*, 335 (1958)
57) Miller, K. J., Ruedenberg, K.: J. Chem. Phys. *48*, 3418 (1968)
58) Van der Velde, G. A., Nieuwpoort, W. C.: Chem. Phys. Letters *13*, 409 (1972)
59) Moffat, J. B.: J. Mol. Structure *15*, 325 (1973)
60) Pamuk, H. Ö.: Theoret. Chim. Acta *28*, 85 (1972)
61) Sinanoğlu, O., Pamuk, H. Ö.: J. Amer. Chem. Soc. *95*, 5435 (1973)
62) Clementi, E., Popkie, H.: J. Chem. Phys. *57*, 4870 (1972)
63) Lie, G. C., Clementi, E.: J. Chem. Phys. *60*, 1275 (1974)
64) Wigner, E.: Phys. Rev. *46*, 1002 (1934)
65) Kistenmacher, H., Popkie, H., Clementi, E., Watts, R. O.: J. Chem. Phys. *60*, 4455 (1974)
66) Kistenmacher, H., Popkie, H., Clementi, E.: J. Chem. Phys. *59*, 5842 (1973)
67) Clementi, E.: Proc. Nat. Acad. Sci. USA *69*, 2942 (1972)
68) Lie, G. C., Clementi, E.: J. Chem. Phys. *60*, 1288 (1974)
69) Watson, R. E.: Phys. Rev. *119*, 170 (1960)
70) Sinanoğlu, O., Pamuk, H. Ö.: Theoret. Chim. Acta *27*, 289 (1972)
71) Duben, A. J., Goddman, L., Pamuk, H. Ö., Sinanoğlu, O.: Theoret. Chim. Acta *30*, 177 (1973)
72) Pamuk, H. Ö.: J. Amer. Chem. Soc. *98*, 7948 (1976)
73) Čársky, P., Zahradník, R., Hobza, P.: Theoret. Chim. Acta *40*, 287 (1975)
74) Koller, J., Šolmajer, T., Borštnik, B., Ažman, A.: J. Mol. Structure *26*, 439 (1975)
75) Hobza, P., Čársky, P., Zahradník, R.: Collect. Czech. Chem. Commun., to be published
76) Nesbet, R. K.: Adv. Chem. Phys. *14*, 1 (1969)
77) Jungen, M., Ahlrichs, R.: Theoret. Chim. Acta *17*, 339 (1970)
78) Ahlrichs, R., Lischka, H., Staemmler, V., Kutzelnigg, W.: J. Chem. Phys. *62*, 1225 (1975)
79) Ahlrichs, R., Lischka, H., Zurawski, B., Kutzelnigg, W.: J. Chem. Phys. *63*, 4685 (1975)
80) Gelus, M., Ahlrichs, R., Staemmler, V., Kutzelnigg, W.: Theoret. Chim. Acta *21*, 63 (1971)
81) Gelus, M., Kutzelnigg, W.: Theoret. Chim. Acta *28*, 103 (1973)
82) Lischka, H.: Theoret. Chim. Acta *31*, 39 (1973)
83) Staemmler, V.: Theoret. Chim. Acta *31*, 49 (1973)
84) Kutzelnigg, W., Staemmler, V., Hoheisel, C.: Chem. Phys. *1*, 27 (1973)
85) Tsupline, B., Kutzelnigg, W.: Chem. Phys. Letters *23*, 173 (1973)
86) Lischka, H.: Chem. Phys. Letters *20*, 448 (1973)
87) Lischka, H.: Chem. Phys. *2*, 191 (1973)
88) Lischka, H.: J. Amer. Chem. Soc. *96*, 4761 (1974)
89) Grimaldi, F., Lecourt, A., Moser, C.: Intern. J. Quantum Chem. *S1*, 153 (1967)
90) Bunge, C. F.: Phys. Rev. *168*, 92 (1968)
91) Langhoff, S. R., Davidson, E. R.: Intern. J. Quantum Chem. *8*, 61 (1974)
92) Yoshimine, M.: J. Computational Phys. *11*, 449 (1973)
93) Roos, B.: Chem. Phys. Letters *15*, 153 (1972)
94) Buenker, R. J., Peyerimhoff, S. D.: Theoret. Chim. Acta *39*, 217 (1975)
95) Bender, C. F., Davidson, E. R.: J. Phys. Chem. *70*, 2675 (1966)
96) Schaefer III, H. F., Bender, C. F.: J. Chem. Phys. *55*, 1720 (1971)
97) Bender, C. F., Schaefer III, H. F.: J. Chem. Phys. *55*, 4798 (1971)

98) Meyer, W.: Intern. J. Quantum Chem. *S5*, 341 (1971)
99) Wahl, A. C., Das, G.: Advan. Quantum Chem. *5*, 261 (1970)
100) Hinze, J.: J. Chem. Phys. *59*, 6424 (1973)
101) Clementi, E., Veillard, A.: J. Chem. Phys. *44*, 3050 (1966)
102) Gregory, A. R.: Chem. Phys. Letters *11*, 271 (1971)
103) Gregory, A. R., Paddon-Row, M. N.: Chem. Phys. Letters *12*, 552 (1972)
104) Meyer, W.: J. Chem. Phys. *58*, 1017 (1973)
105) Kelly, H. P.: Phys. Rev. *A134*, 1450 (1964)
106) Meyer, W.: Theoret. Chim. Acta *35*, 277 (1974)
107) Meyer, W., Rosmus, P.: J. Chem. Phys. *63*, 2356 (1975)
108) Rosmus, P., Meyer, W.: J. Chem. Phys. *66*, 13 (1977)
109) Ahlrichs, R., Keil, F., Lischka, H., Kutzelnigg, W., Staemmler, V.: J. Chem. Phys. *63*, 455 (1975)
110) Mehler, E. L.: Theoret. Chim. Acta *35*, 17 (1974)
111) Bartlett, R. J., Silver, D. M.: J. Chem. Phys. *62*, 3258 (1975)
112) Bartlett, R. J., Silver, D. M.: J. Chem. Phys. *64*, 4578 (1976)
113) Silver, D. M., Bartlett, R. J.: Phys. Rev. *A13*, 1 (1976)
114) Freemen, D. L., Karplus, M.: J. Chem. Phys. *64*, 2641 (1976)
115) Pople, J. A., Binkley, J. S., Seeger, R.: Intern. J. Quantum Chem. *S10*, 1 (1976)
116) Urban, M., Kellö, V., Hubač, I.: Chem. Phys. Letters *51*, 170 (1977)
117) Mehler, E. L., van der Velde, G. A., Nieuwpoort, W. C.: Intern. J. Quantum Chem. *S9*, 245 (1975)
118) Hosteny, R. P., Gilman, R. R., Dunning, Jr., T. H., Pipano, A., Shavitt, I.: Chem. Phys. Letters *7*, 325 (1970)
119) Pipano, A., Shavitt, I.: Intern. J. Quantum Chem. *2*, 741 (1968)
120) Cederbaum, L. S., Schönhammer, K., von Niessen, W.: Chem. Phys. Letters *34*, 392 (1975)
121) Goldhammer, P., Feenberg, E.: Phys. Rev. *101*, 1233 (1956)
122) Diercksen, G. H. F., Sutcliffe, B. T.: Theoret. Chim. Acta *34*, 105 (1974)
123) Schönhammer, K., Cederbaum, L. S.: Phys. Letters *A51*, 325 (1975)
124) Cederbaum, L. S., Schönhammer, K.: Phys. Rev. *A12*, 2257 (1975)
125) Ref.[5], p. 206
126) Das, G., Wahl, A. C.: Phys. Rev. Letters *24*, 440 (1970)
127) Cederbaum, L. S., Hohlneicher, G., v. Niessen, W.: Mol. Phys. *26*, 1405 (1973)
128) Cederbaum, L. S., v. Niessen, W.: Chem. Phys. Letters *24*, 263 (1974)
129) Hubač, I., Kvasnička, V., Holubec, A.: Chem. Phys. Letters *23*, 381 (1973)
130) Hubač, I., Urban, M.: Theoret. Chim. Acta *45*, 185 (1977)
131) Cederbaum, L. S.: J. Phys. B: Atom. Mol. Phys. *8*, 290 (1975)
132) Kaldor, U.: Phys. Rev. Letters *31*, 1338 (1973)
133) Kaldor, U.: J. Chem. Phys. *63*, 2199 (1975)
134) Stern, P. S., Kaldor, U.: J. Chem. Phys. *64*, 2002 (1976)
135) Hubač, I., Liška, M.: to be published
136) Hubač, I., Kvasnička, V.: Croat. Chem. Acta, in press
137) Löwdin, P. O.: Adv. Phys. *5*, 1 (1956)
138) Kvasnička, V., Laurinc, V., Hubač, I.: Phys. Rev. *A10*, 2016 (1974)
139) Laurinc, V., Hubač, I., Kvasnička, V.: Theoret. Chim. Acta *32*, 265 (1974)

Received July 28, 1977

# A Quantitative Measure of Chemical Chirality and Its Application to Asymmetric Synthesis

**James Dugundji**

Department of Mathematics, University of Southern California, Los Angeles, California 90007, U.S.A.

**Rosemarie Kopp, Dieter Marquarding, and Ivar Ugi**

Organisch-Chemisches Institut der Technischen Universität München, Lichtenbergstraße 4, 8046 Garching, Germany

A notion of chirality, called chemical chirality, is presented, which is meaningful even for non-rigid molecules and for ensembles of such molecules. The concept of chiral genus is then introduced as a quantitative objective measure for chemical chirality. By comparing the genera of the reactants and the final products one can determine whether or not chemical chirality has increased during a reaction. This leads to a precise definition of "asymmetric synthesis" as a stereoselective reaction in which the increase of chirality is due to chiral influences only; numerous examples are given to show that use of this definition gives results that agree with generally accepted notions about these syntheses. After a discussion of stereoselective reactions a general scheme for constructing asymmetric syntheses having high overall yield of the desired product with unlimited isomer purity is proposed. Finally, the definition is compared to a description of asymmetric syntheses which is valid only under restrictive conditions.

## Table of Contents

## A. Introduction

In modern stereochemistry asymmetric reactions[1–24] play an increasing rôle in the solution of problems involving chiral chemical compounds and in their preparation. Asymmetric reactions are useful in elucidating the structural features of chiral molecules and in the study of reaction mechanisms, including those of biochemical reactions. Their importance is further emphasized by the realization that all naturally occurring chiral chemical compounds stem from asymmetric syntheses.

The characteristic features of asymmetric syntheses are, intuitively, a stereoselectivity which is due to chiral influences alone and the formation of products which are "more chiral" than the starting materials. No precise and universally applicable description of "more chiral" is available[25] and neither is there any objective criterion for characterizing reactions whose stereoselectivity is due to chiral influences alone.

Our objective in this paper is to present precise descriptions for these concepts and thereby attain an objective general definition of "asymmetric synthesis". We illustrate use of this definition in a broad variety of stereoselective reactions; those reactions generally counted to be asymmetric do in fact turn out to be asymmetric by our criterion, and some of those whose asymmetry is controversial do not constitute asymmetric syntheses by our definition.

## B. Chirality

The original terms "molecular asymmetry" and "molecular dissymmetry" in classical stereochemistry were gradually replaced by Lord Kelvin's geometric concept of chirality[26] as it was realized that this concept was more descriptive of the actual chemical structures. However, his concept pertains only to rigid models, and recent developments in chemistry have shown that the existence of a rigid model for some or all of the geometric arrangements of a molecule is not relevant for chirality: one must also account for the chirality of molecules which are known to be so flexible that they cannot be represented by any single rigid model[25, 27]. Therefore, we propose a broader definition of molecular chirality, which appears to be more representative of the currently known facts in chemistry.

*Definition:* A molecule is chemically chiral under given observation conditions if there is one momentary geometry of that molecule, which cannot be superimposed on its mirror image by using only rotations, translations, and those intramolecular motions that can occur under the observation conditions.

Note that this differs from Lord Kelvin's notion in that it involves the observation conditions and the intramolecular motions that can occur under those conditions. It is known, for example, that under suitable conditions, some molecules which cannot be superimposed on their mirror image by any combination of rotations and translations may nevertheless convert to their mirror image by intramolecular motions alone; such molecules cannot be observed as chiral under the stated conditions. In particular, this definition of chemical chirality gives meaning to the concept of an "enantiomer pair" which is more in keeping with its usage in chemistry: the enantiomers may be stable molecular individuals, or be interconverting under the given observation conditions.

This definition is well-suited for the generation and discussion of chirality functions for flexible molecules, as will be shown in a subsequent paper[28].

*Examples*

a) Since any conformation of *meso*-tartaric acid *1* is interconvertible with its enantiomer (*e.g.* $1a \rightleftharpoons \overline{1a}$) by intramolecular motions, *1* (and all of its conformations) is chemically achiral, although all of its conformations, except *1b* and *1d*, are geometrically chiral. Note that the geometric achirality of *1b* and *1d* is not a prerequisite for the chemical achirality of *1*.

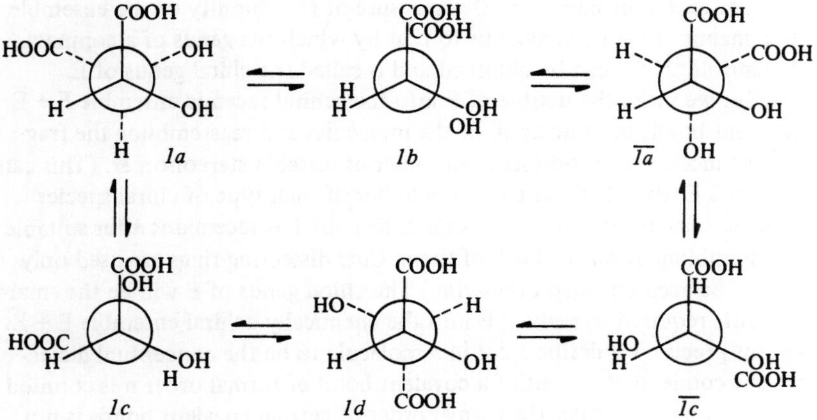


b) The conformations of *2* are geometrically chiral but chemically achiral, because they are all interconvertible with the respective enantiomers through internal rotations of the type $2 \rightleftharpoons \overline{2}$[29,30].

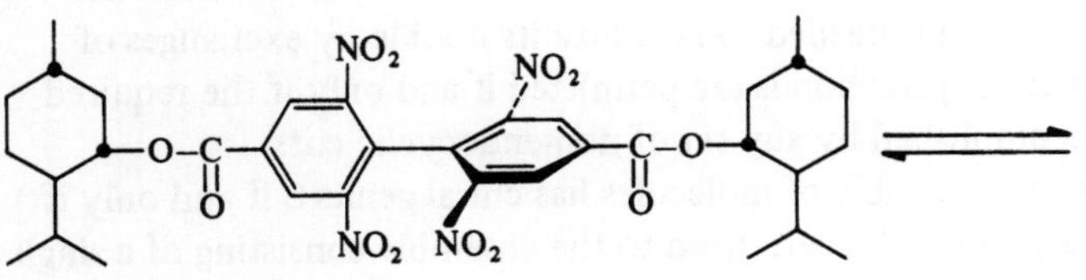


*2*

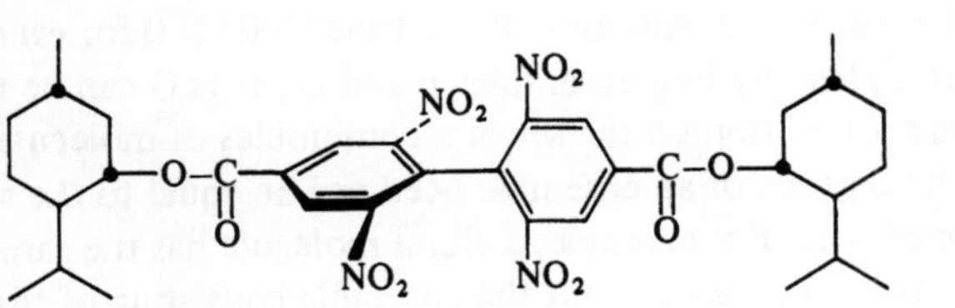


$\overline{2}$

## C. Chiral Genus

Using the concept of chirality as defined above, we follow the program stated in our introduction and now aim to provide a quantitative, objective means for expressing the chirality of an ensemble of molecules. The fundamental notion is that of a chemically achiral ensemble of molecules:

*Definition:* An ensemble of molecules is chemically achiral under given observation conditions if and only if it is composed of racemic pairs of enantiomers and achiral molecules.

The double of a given ensemble E of molecules is the ensemble 2 E which contains two copies of each molecule of E. Our measure of the chirality of the ensemble E is gotten in a manner roughly analogous to that by which the genus of a compact orientable 2-manifold[31–33] can be obtained and is called the chiral genus of E. Stated informally: we make the double of E into the achiral racemic ensemble $E + \overline{E}$ by cutting covalent bonds in some or all of the molecules and reassembling the fragments of each cut molecule to form its enantiomer or suitable stereoisomer. (This can always be done in 2 E, since 2 E has an even number of each type of chiral species found in E, and each such pair can, for example, be paired as racemates after suitable cutting and reassembling of one or both of them. Cuts dissecting rings are used only if this end cannot be accomplished otherwise.) The chiral genus of E will be the smallest number of cuts required to make 2 E into the chemically achiral ensemble $E + \overline{E}$.

To make this precise, we define a cut in a molecule to be the conceptual dissection of a covalent connectivity; a cut of a covalent bond of formal order n is counted as being n cuts. In some molecules the formal order of certain covalent bonds is not an integer. In such cases one can either approximate the bond order by an integer, or use a fractional bond order consistently for counting the cuts; the genus for such molecules may therefore be not an integer in the latter case. With these concepts, we can make the precise

*Definition:* The chiral genus of an ensemble of molecules is one-half the minimal number of cuts of covalent bonds needed to racemize its double by exchanges of molecular parts. Cuts of endocyclic bonds are permitted if and only if the required conversion cannot be accomplished by any set of non-endocyclic cuts.

With this definition, an ensemble of molecules has chiral genus 0 if and only if it is chemically achiral. Applying the definition to the ensemble consisting of a single molecule, we obtain that the chiral genus of a single molecule is 1/2 the number of cuts which are necessary to convert it into its enantiomer. More generally, letting $G(E)$ denote the chiral genus of any ensemble E, we have $G(E) \geqslant 0$ for every E and $G(E \cup E') \leqslant G(E) + G(E')$ for any two ensembles E and E′; thus G can be regarded as a finitely subadditive set function on the set of all ensembles of molecules.

Observe that the chiral genus of an ensemble need not be equal to the sum of the chiral genus of its members. For example, a chiral molecule has the same (non-zero) chiral genus as its enantiomer, but the ensemble consisting of that molecule and its enantiomer has chiral genus zero.

*Examples*

a) The ensemble [*3* + *3*] has chiral genus 2, because two cuts are needed to convert it to the chemically achiral (racemic) ensemble [*3* + $\overline{3}$]. The ensemble [*3*] consisting of *3* alone has chiral genus 1:

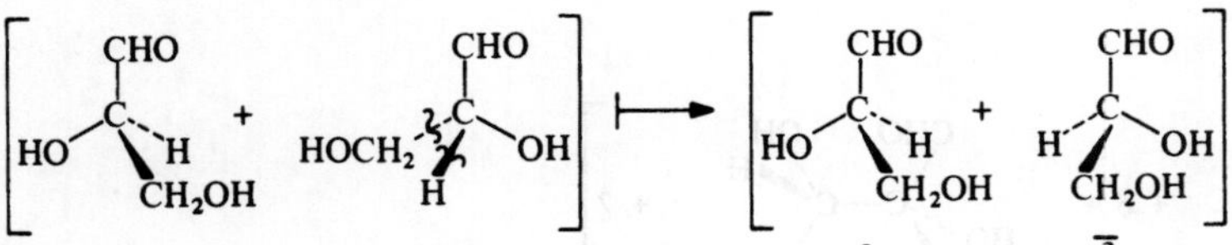

*3* *3* *3* $\overline{3}$

b) The chiral genus of the ensemble [*4* + *4*] is 4, although *4* could be converted into its enantiomer $\overline{4}$ through a pair of endocyclic cuts. These are, however, not permissible, since $4 \longrightarrow \overline{4}$ can be achieved without dissecting rings:

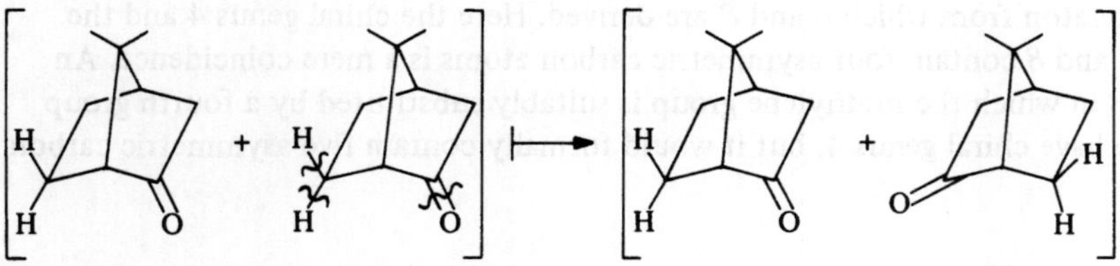

*4* *4* *4* $\overline{4}$

c) Inositol *5* with its six asymmetric carbon atoms has genus 2, because the double ensemble [*5* + *5*] has genus 4. Note that, as remarked, the genus of a molecule is one-half the minimal number of cuts required to convert it to its enantiomer:

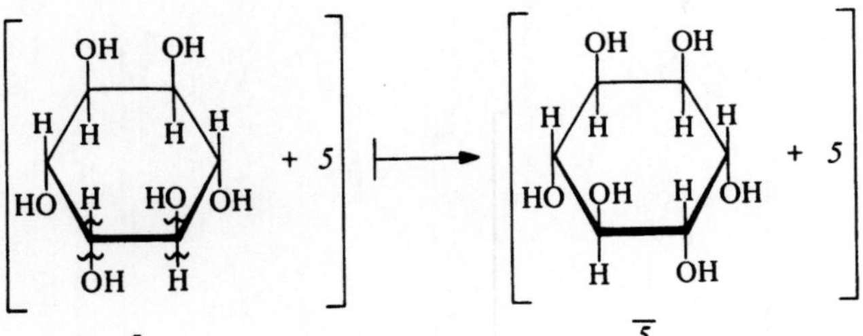

*5* $\overline{5}$

d) The chiral genus of the ensemble [*6* + *6* + *7* + *7*] is 4. The chiral genus of [*6* + *6*] is also 4, whereas the chiral genus of [*7* + *7*] is only 2. Note that

$$G([6+6+7+7]) < G([6+6]) + G([7+7])$$

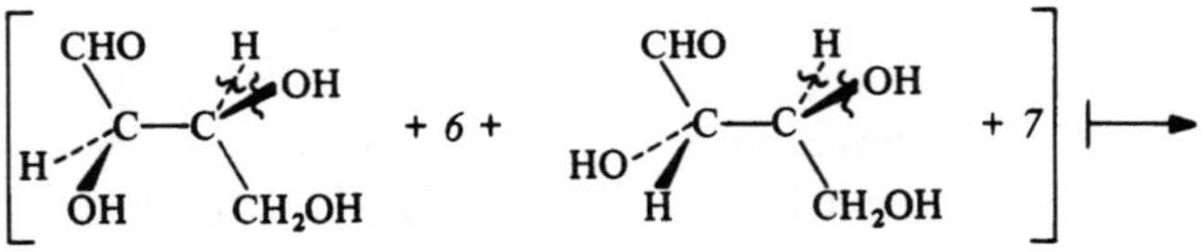


*6* *7*

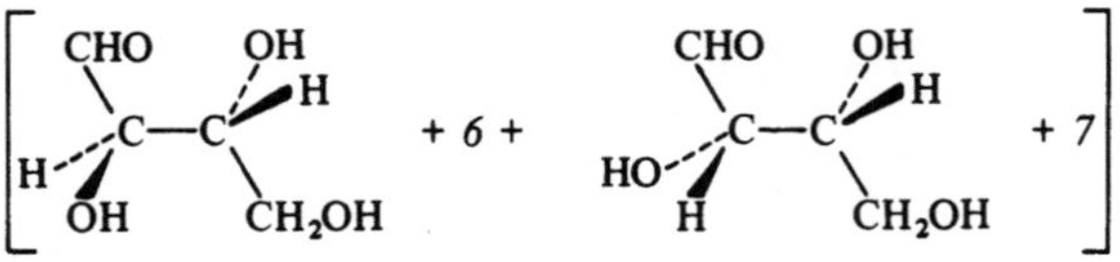


$\overline{7}$ $\overline{6}$

e) The chiral genus of the ensemble [*8* + *8*] is 4. Note that racemization of [*8* + *8*] by a minimum number of cuts is associated with a permutation of molecular parts according to a pattern corresponding to a fourfold improper rotation of the spiropentane skeleton from which *8* and $\overline{8}$ are derived. Here the chiral genus 4 and the fact that *8* and $\overline{8}$ contain four asymmetric carbon atoms is a mere coincidence. An analog of *8* in which the methylene group is suitably substituted by a fourth group would also have chiral genus 4, but it would formally contain five asymmetric carbon atoms:

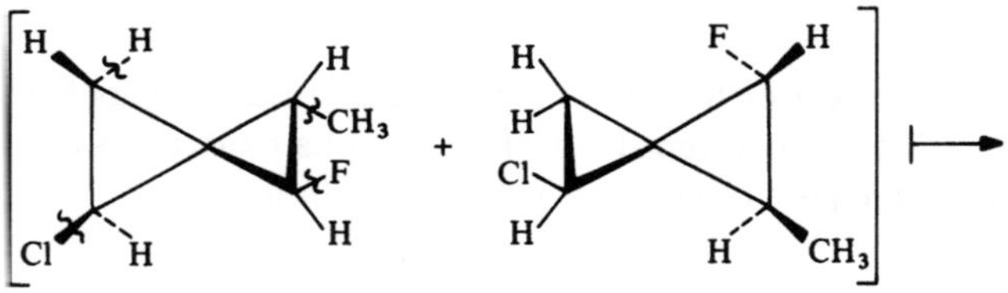


*8* *8*

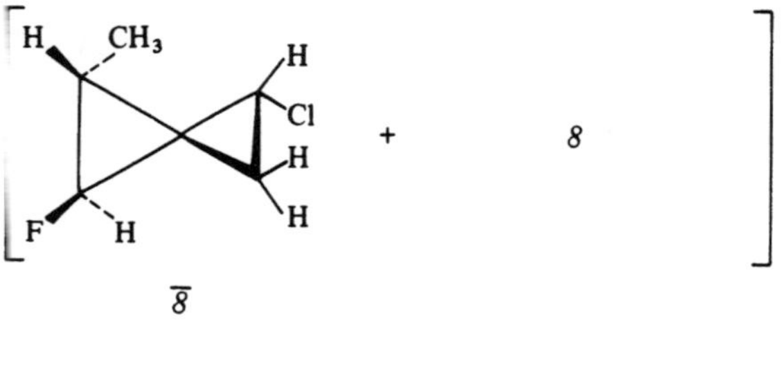


$\overline{8}$

f) The determination of chiral genus of longifolene *9*, twistane *10*, tri-o-thymotide *11*, and 6,6′-dinitrodiphenic acid *12* is indicated below.

Note that endocyclic bonds must be cut in compounds *9*–*11*.

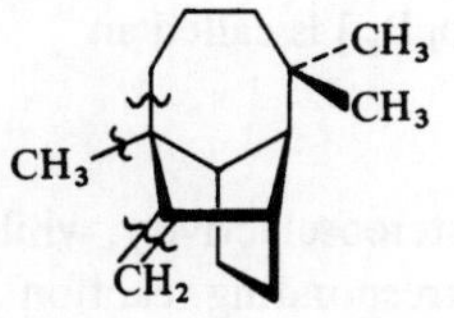

9

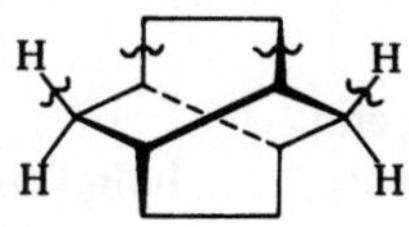

10

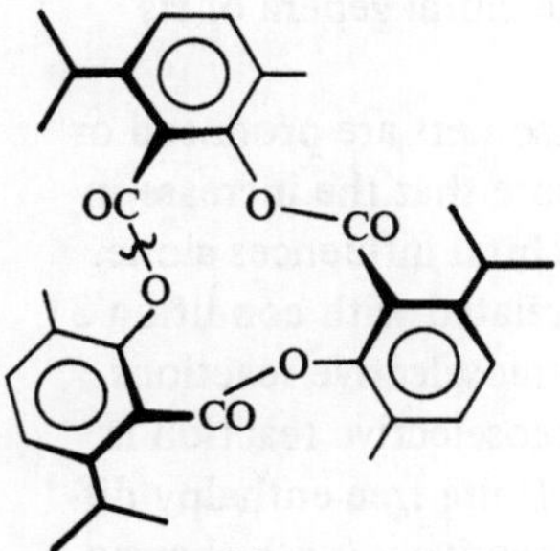

11

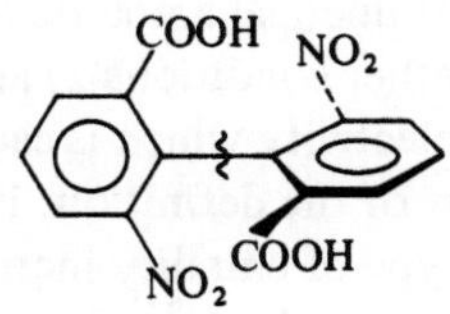

12

## D. Asymmetric Syntheses and Stereoselectivity

Since a quantitative measure for chirality is now available, we apply it to the study of chemical reactions. We consider the ensemble of the initial reactants and the ensemble of the final products; in both ensembles each molecule is taken as many times as it appears in the balanced stoichiometric equation representing the reaction. These are called the ensembles representing the initial reactants and the final products. With this convention, we can now divide all chemical reactions involving chiral molecules into three mutually exclusive categories:

1. Chirality preserving reactions: those for which the ensemble representing the initial reactants and the final products both have the same chiral genus.

2. Chirality dissipating reactions: those for which the chiral genus of the ensemble representing the final products is smaller than that of the ensemble representing the initial reactants.

3. Chirality increasing reactions: those for which the chiral genus of the ensemble representing the final products is greater than that of the ensemble representing the initial reactants.

If $A_1, \ldots, A_n$ are any molecules then in the balanced stoichiometric equation $[a_1A_1, \ldots, a_nA_n]$ denotes the ensemble of initial reactants that contains $a_1$ copies of $A_1, \ldots,$ and $a_n$ copies of $A_n$, analogous considerations apply to the final products $B_1, \ldots, B_k$.

*Definition:* A reaction $[a_1A_1 + \ldots + a_nA_n] \longrightarrow [b_1B_1 + \ldots + b_kB_k]$ is called an asymmetric synthesis, if

1. the reaction is stereoselective, and
2. $G([a_1A_1, \ldots, a_nA_n]) < G([b_1B_1, \ldots, b_kB_k])$ is valid with stereoselectivity, while $G([a_1A_1, \ldots, a_nA_n]) = G([b_1B_1, \ldots, b_kB_k])$ holds for the corresponding reaction without stereoselectivity, and
3. $G([b_1B_1, \ldots, b_kB_k]) < \sum_{i=1}^{k} G([b_iB_i])$, in words, the chiral genus of the ensemble of final products must be strictly smaller than the sum of the chiral genera of its individual members.

The condition 1) states that unequal amounts of stereoisomers are produced or destroyed by the reaction; together conditions 2) and 3) assure that the increase in chiral genus is due to a stereoselectivity which is caused by chiral influences alone. Condition 1) is an essential part of the definition: it is interrelated with condition 3), and together they restrict the type of chirality increasing stereoselective reactions that will be called an asymmetric synthesis[34]. In fact, a stereoselective reaction is never "infinitely selective", because this would require an infinite free enthalpy difference of stereoisomers, stereoisomeric transition states respectively (with thermodynamic control, or with productive selectivity); in the case of an idealized destructive selectivity "infinite selectivity" would be reached at infinite reaction time, when none of the considered stereoisomers would be left over[35].

*Examples*

a) Reactions *13* ⟶ *14*[36] and *15* ⟶ *16*[37, 38] are chirality preserving:

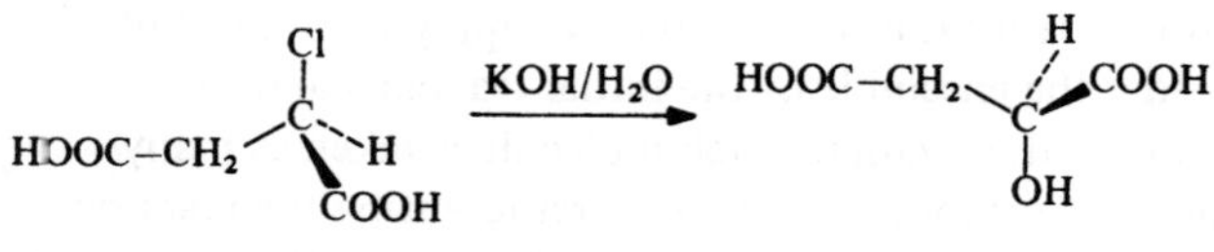


*13* *14*

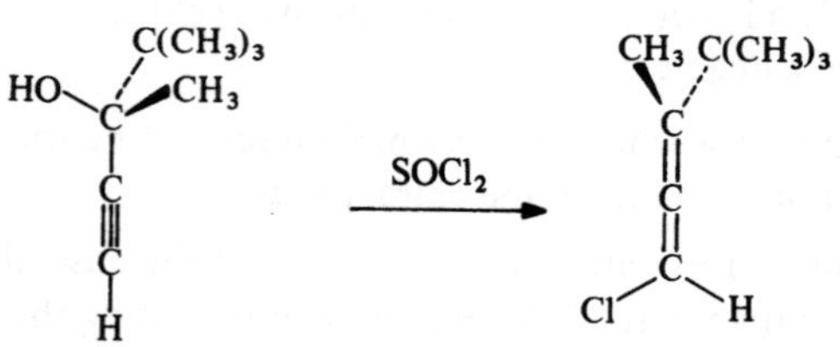


*15* *16*

b) Chirality is preserved for reaction *17* ⟶ *18* for all values of n and m:

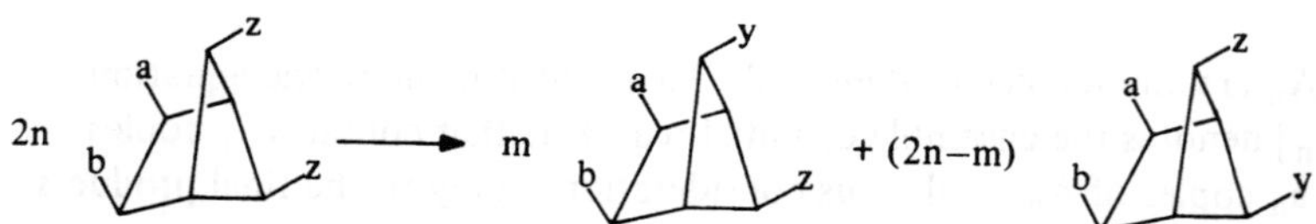


*17* *18a* *18b*

c) Chirality is dissipated by the $S_N1$ reaction *19* ⟶ *20* and also by the reaction *21* ⟶ *22*[24]; $G([2n \cdot 21]) = 4n$, $G([m \cdot 22a + (2n - m) \cdot 22b]) = 2n$.

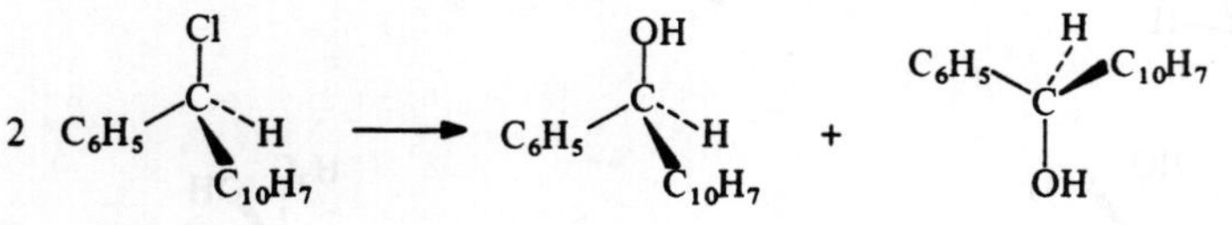

*19* *20* $\overline{20}$

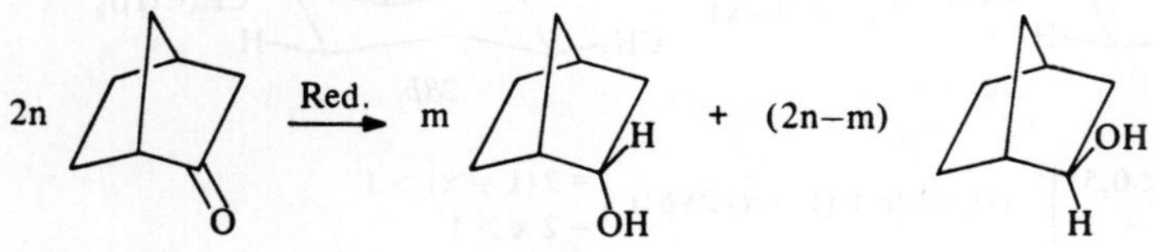

*21* *22a* *22b*

d) An increase of chirality is associated with reactions *22a* ⟶ *21* and *23* ⟶ *24*:

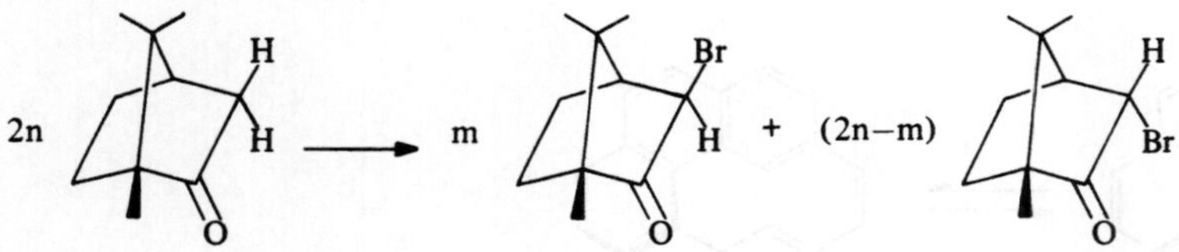

*23* *24a* *24b*

e) The stereoselective decarboxylation of *25* in the presence of brucine as the chiral reference system by Marckwald[2] is one of the first asymmetric syntheses which was carried out *in vitro.*

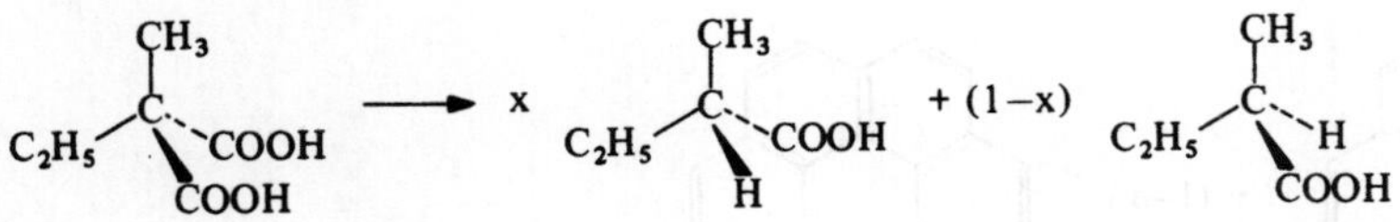

*25* *26* $\overline{26}$

$$G([25]) = 0 \quad \begin{array}{l} (0 < x < 0{,}5)\text{: } G([x26 + (1-x)\,\overline{26}\,]) = 1 - 2x > 0 \\ (0{,}5 < x < 1)\text{: } G([x26 + (1-x)\,\overline{26}\,]) = 2x - 1 > 0 \\ G([x26]) + G([(1-x)\,\overline{26}\,]) = 1 \end{array}$$

f) McKenzie's asymmetric synthesis[3] of *28a* and *28b* by the addition of methyl magnesium iodide to *27* is also one of the first *in vitro* asymmetric syntheses and later studied in the context of Prelog's rule[11, 39, 40].

Note that $G([x28a + (1-x)\,28b]) < xG([28a]) + (1-x)G([28b])$ is due to the fact that the conversion of *28a* into the enantiomer of *28b* requires only two cuts, whereas its enantiomerization involves at least 4 cuts.

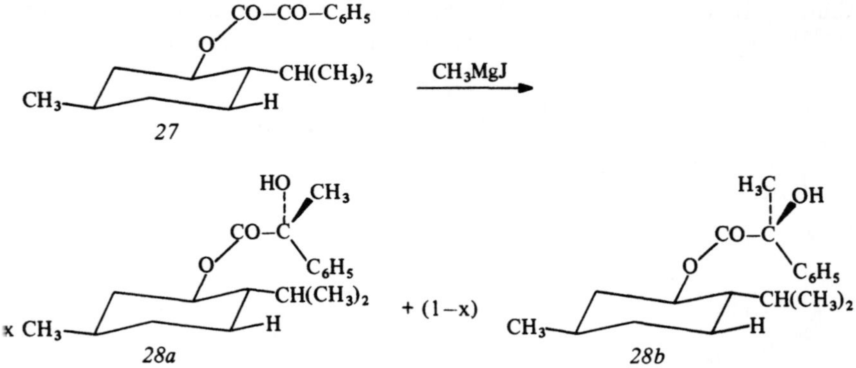

$$G([27]) = 1 \quad \begin{matrix} (0 < x < 0{,}5) \\ (0{,}5 < x < 1) \end{matrix} \Big\} \quad G([x\,28a + (1-x)\,28b]) \Big\{ \begin{matrix} = 2(1-x) > 1 \\ = 2x > 1 \end{matrix}$$

$$xG([28a]) + (1-x)G([28b]) = 2$$

g) The photochemically "absolute asymmetric synthesis"[41–50] of *30* is an asymmetric synthesis according to our definition:

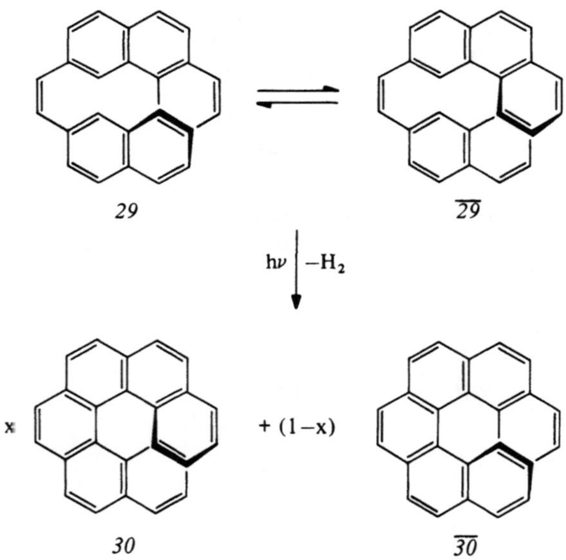

$$G([29 + \overline{29}]) = 0$$

$$\begin{matrix} (0 < x < 0{,}5) \\ (0{,}5 < x < 1) \end{matrix} \quad G([x\,30 + (1-x)\,\overline{30}]) \Big\{ \begin{matrix} = 0{,}5 - x > 0 \\ = x - 0{,}5 > 0 \end{matrix}$$

$$xG([30]) + (1-x)G([\overline{30}]) = 0{,}5$$

h) A stereoselective reaction $31 \longrightarrow x\,32a + (1-x)\,32b$ $(0 < x \neq 0{,}5 < 1)$ belongs according to our definition to the asymmetric syntheses, while the closely related reaction $17 \longrightarrow 18a + 18b$ does not:

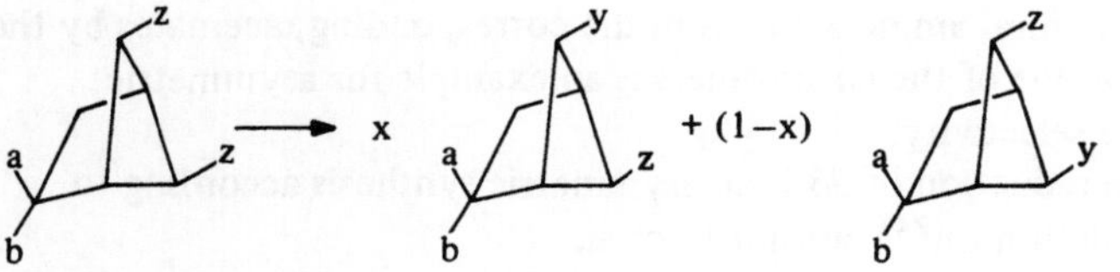

*31* *32a* *32b*

$$G([31]) = 1 \quad \begin{matrix}(0 < x < 0{,}5)\\(0{,}5 < x < 1)\end{matrix}\Big| \quad G([x32a + (1-x)32b]) \begin{cases} = 2(1-x) > 1 \\ = 2x > 1 \end{cases}$$

$$xG([32a]) + (1-x)G([32b]) = 2$$

i) The kinetic resolution of *34* and $\overline{34}$ by *33*[51] qualifies as an asymmetric synthesis only whenever $n > \{^{x}_{1-x}$:

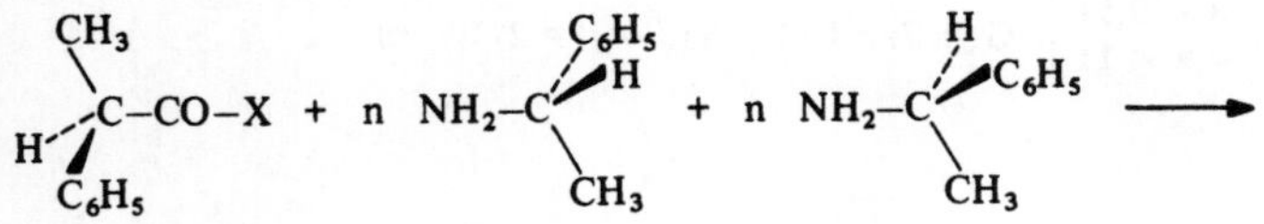

*33* *34* $\overline{34}$

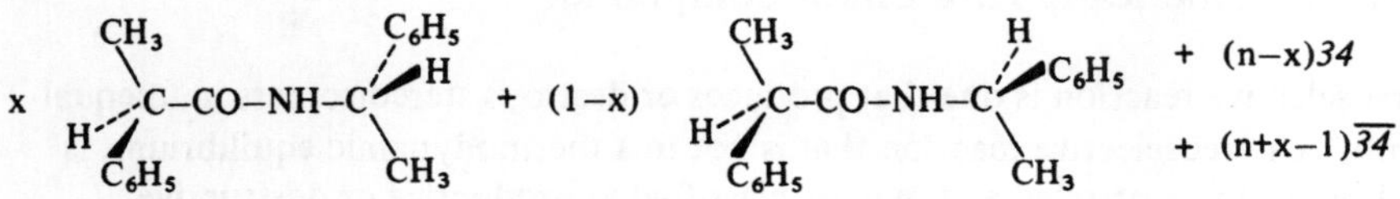

*35a* *35b*

$$G([33 + n34 + n\overline{34}]) = 1$$

$$\begin{matrix}(0 < x < 0{,}5)\\(0{,}5 < x < 1)\end{matrix}\Big| \quad G([x35a + (1-x)35b + (n-x)34 + (n+x-1)\overline{34}]) \begin{cases} = 3 - 4x > 1 \\ = 4x - 1 > 1 \end{cases}$$

$$xG([35a]) + (1-x)G([35b]) + (n-x)G([34]) + (n+x-1)G([\overline{34}]) = 2n + 1;\ (n > \{^{x}_{1-x})$$

j) The stereoselective enzymatic deacetylation of a racemic N-acetylamino acid $(36 + \overline{36})$[52] is an asymmetric synthesis according to our definition as long as $(0 < y < x < 1)$:

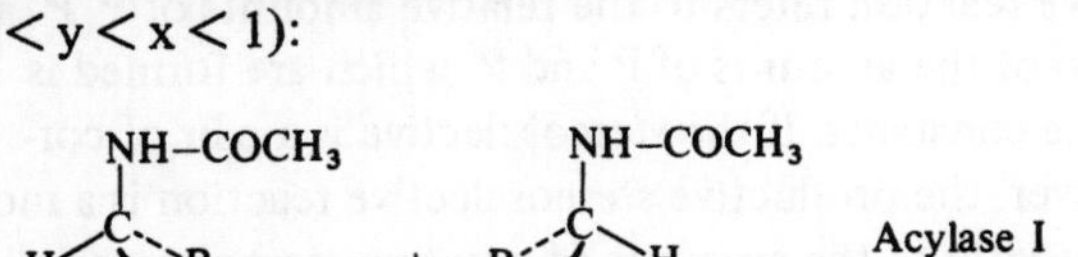

*36* $\overline{36}$

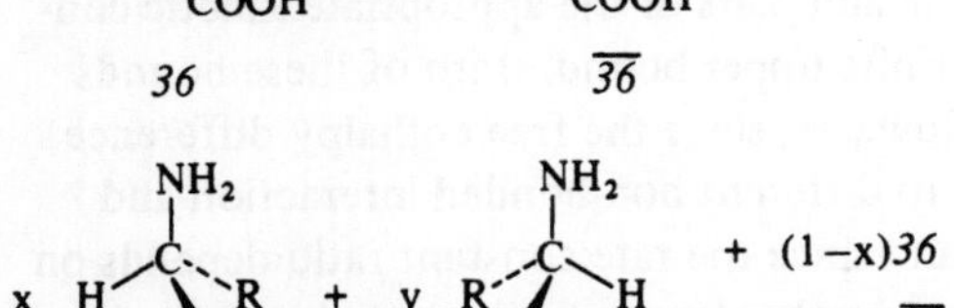

*37* $\overline{37}$

$$G([36 + \overline{36}]) = 0$$

$$G([x-y)37 + (x-y)\overline{36}]) = 2(x-y) > 1, \text{ if } (0 < y < x < 1)$$

$$xG([37]) + yG([\overline{37}]) + (1-x)G([36]) + (1-y)G([\overline{36}]) = 2$$

k) The preparation of chiral amino acids from the corresponding racemates by the selective destruction[52] of one of the enantiomers is an example for asymmetric synthesis with destructive selectivity.

l) The stereoselective reduction of *38* is an asymmetric synthesis according to Mosher's and Morrison's definition[24] but not to ours:

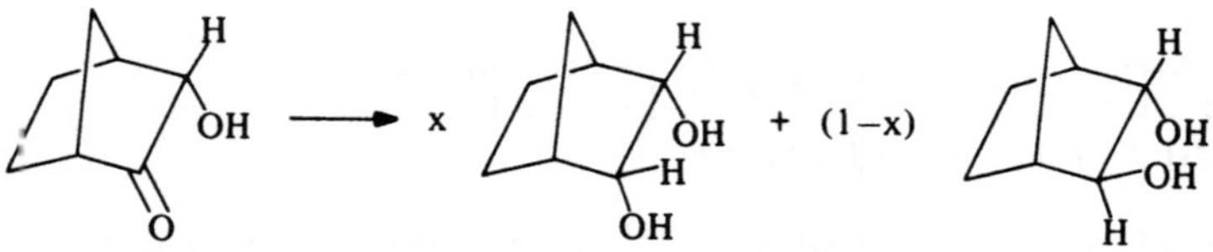

*38* *39a* *39b*

$$G([38]) = 2 \quad \left.\begin{matrix}(0 < x < 0{,}5)\\(0{,}5 < x < 1)\end{matrix}\right\} \quad G([x\,39a + (1-x)\,39b]) = 2(1-x) < 2$$

## E. Productive and Destructive Stereoselective Reactions and the Preparation of Isomerically Pure Chiral Compounds

A stereoselective reaction is one that produces or destroys stereoisomers in unequal amounts. A stereoselective reaction that is not in a thermodynamic equilibrium, is called kinetically controlled and may be classified as productive or destructive.

The productive kinetically controlled stereoselective reactions involve competing parallel reactions which begin with a common ensemble of molecules where all members participate in the formation of the products, as in the scheme

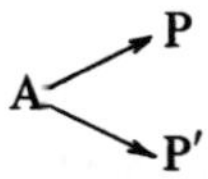

The selectivity of a productive reaction refers to the relative amounts of P, P′ at the time of observation. The ratio of the amounts of P and P′ which are formed is the ratio of the corresponding rate constants, if the stereoselective is a pair of corresponding reactions[53]. If, however, the productive stereoselective reaction is a more complex kinetic scheme, then the ratio of the amounts of any two stereoisomeric products, P and P′, which depends on time and pairs of the appropriate kinetic constants, has a positive lower bound and a finite upper bound. Both of these bounds are the ratios of two rate constants[54]. However, since the free enthalpy difference of stereoisomeric transition states is due to different non-bonded interaction and does not, as a rule, exceed 3 kcal/mole, and since the rate constant ratio depends on the free enthalpy difference, this ratio has a rather low upper bound. Accordingly, the stereoselectivity of productive reactions is generally low (50–90% relative yield of the preferred product in most cases).

The destructive kinetically controlled stereoselective reactions involve independent competing parallel reactions whose initial reactants P, P′ are not identical, as in the scheme

$P \longrightarrow Q$
$P' \longrightarrow Q'$

The stereoselectivity of destructive reactions refers to the relative amounts of initial reactants that remain at the time of observation. Clearly, a destructive reaction is stereoselective if and only if the relative amounts of P and P′ changes with time, since otherwise the rate difference would arise only because of change in the initial concentrations of P and P′. Note that there is no upper bound for the amounts of left-over reactants P and P′ in the destructive reaction[35].

*Example*

The stereoselective four component condensation[55] yields *44*-(S[R]RSS) and *44*-(S[R]SSS) in an isomer ratio of up to 99.5 : 0.5. When stereoisomer mixtures of *44*-(S[R]SSS) and *44*-(S[R]RSS) are subjected to acidolysis *44*-(S[R]SSS) is cleaved much faster than *44*-(S[R]RSS), leading to an enrichment of *44*-(S[R]RSS) up to 99.98% and more. The latter destructive stereoselective reaction may also be regarded as an asymmetric synthesis, since the chirality of the α-ferrocenylisobutyl group of *44* is retained in *46*-(R).

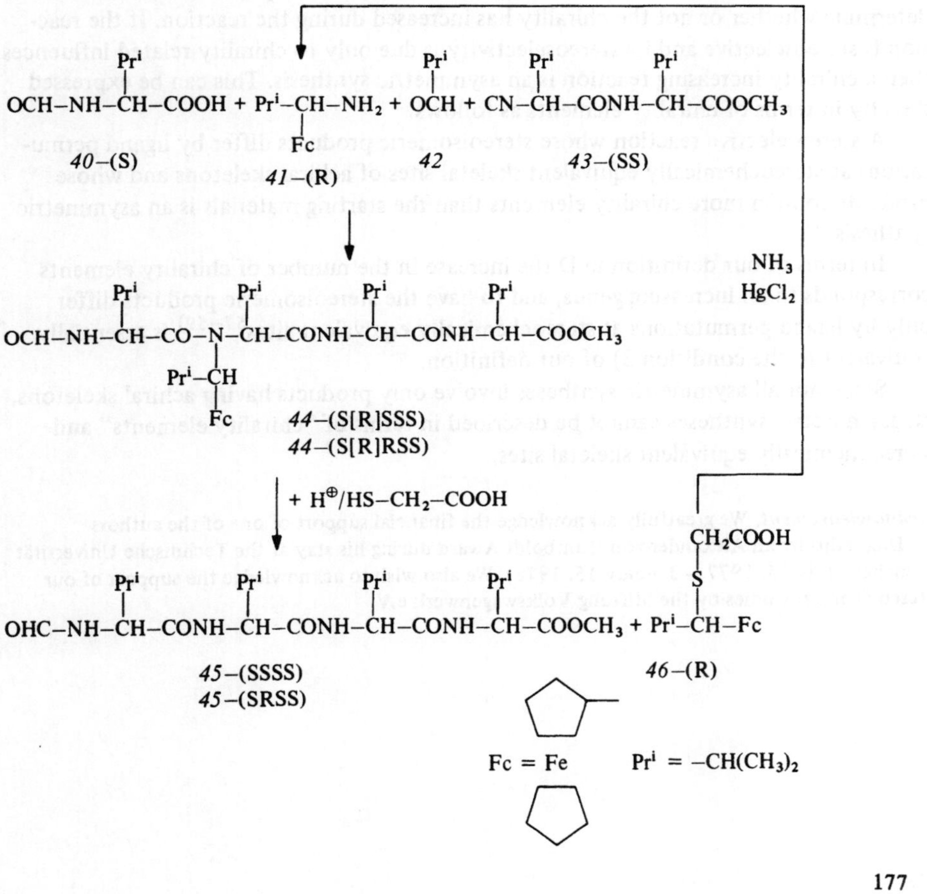

On the basis of this discussion a particularly promising scheme for synthesizing chiral compounds with great isomer purity and high overall yield would be to start with a productive asymmetric synthesis (to assure high yield) and follow by a destructive one (to assure unlimited isomer purity of the desired product). Such a procedure has the essential advantage that under suitable conditions it leads to pure chiral compounds without cumbersome separation of stereoisomers. With proper choice of the reactants the procedure will also permit an effective recovery of the auxiliary chiral materials needed as chiral templates in the asymmetric synthesis.

Predictably, there will be an increasing tendency to prepare chiral chemical compounds by asymmetric syntheses as the understanding of their methodology grows.

## F. Alternative Description of Asymmetric Synthesis for Molecules with a Well-defined Achiral Skeleton

For the special case that all molecules to be considered have well-defined achiral skeletons it is possible to enumerate "chirality elements"[56]. The enumeration of these chirality elements which has been given in our previous paper can be used to determine whether or not the chirality has increased during the reaction. If the reaction is stereoselective and its stereoselectivity is due only to chirality related influences, then a chirality increasing reaction is an asymmetric synthesis. This can be expressed directly in terms of chirality elements as follows:

A stereoselective reaction whose stereoisomeric products differ by ligand permutations at stereochemically equivalent skeletal sites of achiral skeletons and whose products contain more chirality elements than the starting materials is an asymmetric synthesis[57].

In terms of our definition in D the increase in the number of chirality elements corresponds to an increasing genus, and to have the stereoisomeric products differ only by ligand permutations at stereochemically equivalent sites[57–59] is essentially equivalent to the condition 3) of our definition.

Since not all asymmetric syntheses involve only products having achiral skeletons, all asymmetric syntheses cannot be described in terms of "chirality elements" and stereochemically equivalent skeletal sites.

*Acknowledgement.* We greatfully acknowledge the financial support of one of the authors (J. Dugundji) by an Alexander-von-Humboldt Award during his stay at the Technische Universität München (July 15, 1977 – January 15, 1978). We also wish to acknowledge the support of our stereochemical studies by the Stiftung Volkswagenwerk e.V.

## G. References


1) Fischer, E.: Ber. *27*, 3231 (1894)
2) Marckwald, W.: Ber. *37*, 1368 (1904)
3) McKenzie, A.: J. Chem. Soc. (London) *85*, 1249 (1904)
4) McKenzie, A.: Angew. Chem. *45*, 59 (1932)
5) Kortüm, G.: Neuere Forschungen über die optische Aktivität chemischer Moleküle, in: Sammlung Chemie und chemische Technologie. Stuttgart: F. Enke Verlag 1932, Vol. 10
6) Ritchie, P. D.: Asymmetric synthesis and asymmetric induction. London: Oxford University Press 1933
7) Ebel, F.: Asymmetrische Synthese, in: Stereochemie. Freudenberg, K. (ed.). Leipzig und Wien: Franz Deuticke Verlag 1933, p. 580
8) Ritchie, P. D.: Advan. Enzymol. *7*, 65 (1947)
9) Turner, E. E., Harris, M. M.: Quart. Rev. *1*, 299 (1947)
10) Theilacker, W., in: Houben-Weyl, Methoden der organischen Chemie. Müller, E. (ed.). Stuttgart: Thieme Verlag 1955, Vol. 4, Part 2, p. 505
11) Prelog, V.: Bull. Soc. Chim. France *1956*, 987
12) Hirschmann, H.: The structural basis for the differentiation of identical groups in asymmetric reactions, in: Essays in biochemistry. Graff, S. (ed.). New York: John Wiley & Sons, Inc. 1956, p. 156
13) Bláha, K.: Chemie (Prag) *9*, 845 (1957)
14) Eliel, E. L.: Stereochemistry of carbon compounds. New York: McGraw-Hill 1962, p. 81
15) Klabunowski, Je. I.: Asymmetrische Synthese. Berlin: VEB Deutscher Verlag der Wissenschaften 1963
16) Morrison, J. D.: Surv. Progr. Chem. *3*, 147 (1966)
17) Pracejus, H.: Fortschr. chem. Forsch. *8*, 493 (1967)
18) Velluz, L., Valls, J., Mathieu, J.: Angew. Chem. *79*, 774 (1967); Angew. Chem. Internat. Edit. *6*, 778 (1967)
19) Amariglio, A., Amariglio, H., Duval, X.: Ann. Chim. *3*, 5 (1968)
20) Boyd, D. R., McKervey, M. A.: Quart. Rev. *22*, 95 (1968)
21) Mathieu, J., Weill-Raynal, J.: Bull. Soc. Chim. France *1968*, 1211
22) Izumi, Y.: Angew. Chem. *83*, 956 (1971); Angew. Chem. Internat. Edit. *10*, 871 (1971)
23) Bähr, W., Theobald, H.: Organische Stereochemie. Begriffe und Definitionen. Berlin-Heidelberg-New York: Springer 1973, p. 8
24) Morrison, J. D., Mosher, H. S.: Asymmetric reactions. Washington, D. C.: American Chemical Society 1976
25) Mislow, K., Bickart, P.: Israel J. Chem. *15*, 1 (1976/77)
26) Lord Kelvin: Baltimore lectures. London: Clay, C. J. and Sons 1904
27) Eliel, E. L.: Israel J. Chem. *15*, 7 (1976/77)
28) Dugundji, J., Kopp, R., Marquarding, D., Ugi, I.: in preparation
29) Mislow, K.: Science *120*, 232 (1954)
30) Mislow, K.: Trans. N. Y. Acad. Sci. *14*, 298 (1957)
31) Lifschitz, S.: Introduction to topology. Princeton: Princeton University Press 1949
32) Lietzmann, W.: Anschauliche Topologie. München: Oldenbourg 1955
33) Arnold, B. H.: Elementare Topologie. Göttingen: Vandenhoeck & Ruprecht 1971
34) In an asymmetric synthesis the effect of a chiral reference system is due to the fact that it interacts with a chiral molecular system and the mirror image of the latter in a different manner, just like a right hand interacts differently with right and left-handed gloves. These differences in interaction are chemically observable as stereoselectivity. Complex schemes of chemical reactions may contain asymmetric syntheses as partial reactions which can be conceptually separated from the other parts of the scheme. In some of these cases the whole reaction scheme may even not qualify as an asymmetric synthesis, because it leads to dissipation of chirality due to the presence of chirality dissipating components whose contribution to the overall result is predominating
35) Brandt, J., Jochum, C., Ugi, I., Jochum, P.: Tetrahedron *33*, 1353 (1977)
36) Walden, P.: Ber. *29*, 133 (1896); *30*, 3146 (1897)

37) Evans, R. J. D., Landor, S. R., Taylor Smith, R.: J. Chem. Soc. *1963,* 1506
38) Evans, R. J. D., Landor, S. R.: J. Chem. Soc. *1965,* 2553
39) Prelog, V.: Helv. chim. Acta *36,* 308 (1953); see also Ref.[40]
40) Cram, D. J., Abd Elhafez, F. A.: J. Amer. Chem. Soc. *74,* 5828 (1952)
41) Moradpour, A., Nicoud, J. F., Balavoine, G., Kagan, H., Tsoucaris, G.: J. Amer. Chem. Soc. *93,* 2353 (1971)
42) Tsoucaris, G., Balavoine, G., Moradpour, A., Nicoud, J. F., Kagan, H.: Compt. Rend. (B) *272,* 1271 (1971)
43) Nicoud, J. F., Kagan, H. B.: Israel J. Chem. *15,* 78 (1976/77)
44) Kagan, H., Moradpour, A., Nicoud, J. F., Balavoine, G., Martin, R. H., Cosyn, J. P.: Tetrahedron Lett. *1971,* 2479
45) Martin, R. H., Schurter, J. J.: Tetrahedron *28,* 1749 (1972)
46) Cochez, Y., Martin, R. H., Jespers, J.: Israel J. Chem. *15,* 29 (1976/77)
47) Bernstein, W. J., Calvin, M., Buchardt, O.: Tetrahedron Lett. *1972,* 2195
48) Bernstein, W. J., Calvin, M., Buchardt, O.: J. Amer. Chem. Soc. *94,* 494 (1972)
49) Bernstein, W. J., Calvin, M., Buchardt, O.: J. Amer. Chem. Soc. *95,* 527 (1973)
50) Buchardt, O.: Angew. Chem. *86,* 222 (1974); Angew. Chem. Internat. Edit. *13,* 179 (1974)
51) Herlinger, H., Kleimann, H., Ugi, I.: Liebigs Ann. Chem. *706,* 37 (1967)
52) Greenstein, J. P., Birnbaum, S. M., Levintow, L.: Biochem. Prep. *3,* 84 (1953)
53) Ruch, E., Ugi, I.: Topics Stereochem. *4,* 99 (1969)
54) Ugi, I., Kaufhold, G.: Liebigs Ann. Chem. *709,* 11 (1967)
55) Urban, R., Eberle, G., Marquarding, D., Rehn, D., Rehn, H., Ugi, I.: Angew. Chem. *88,* 644 (1976); Angew. Chem. Internat. Edit. *15,* 627 (1976)
56) Gasteiger, J., Gillespie, P., Marquarding, D., Ugi, I.: Topics Curr. Chem. *48,* 1 (1974)
57) Gasteiger, J., Marquarding, D., Ugi, I., in: Handbook of stereochemistry, Kagan, H. (ed.). Stuttgart: Thieme Verlag in press
58) Dugundji, J., Marquarding, D., Ugi, I.: Chem. Scripta *9,* 74 (1976)
59) Dugundji, J., Marquarding, D., Ugi, I.: Chem. Scripta *11,* 17 (1977)

Received January 5, 1978

# Author Index Volumes 26–75

*The volume numbers are printed in italics*

# Topics in Current Chemistry

Fortschritte der chemischen Forschung
Managing Editor: F. L. Boschke

Volume 62
**Synthetic and Mechanistic Organic Chemistry**
1976. 14 figures, 63 tables. IV, 258 pages
ISBN 3-540-07525-9

Volume 63
**Bonding and Structure**
1976. 55 figures, 25 tables. IV, 202 pages
ISBN 3-540-07605-0

Volume 64
**Inorganic Biochemistry**
1976. 85 figures. IV, 225 pages
ISBN 3-540-07636-0

Volume 65
**Theoretical Inorganic Chemistry II**
1976. 47 figures, 44 tables. IV, 153 pages
ISBN 3-540-07637-9

Volume 66
**Triplet States 3**
1976. 10 figures. IV, 154 pages
ISBN 3-540-07655-7

Volume 67
**Preparative Organic Chemistry**
1976. 4 figures. IV, 136 pages
ISBN 3-540-07900-9

Volume 68
**Theory**
1976. 36 figures. IV, 156 pages
ISBN 3-540-07932-7

Volume 69
**Inorganic Biochemistry II**
1977. 86 figures, 25 tables. IV, 204 pages
ISBN 3-540-08157-7

Volume 70
**Structural Theory of Organic Chemistry**
1977. 60 figures, 58 tables. VIII, 250 pages
ISBN 3-540-08099-6

Volume 71
**Inorganic Chemistry. Metal Carbonyl Chemistry**
1977. 51 figures, 54 tables. IV, 190 pages
ISBN 3-540-08290-5

Volume 72
**Medicinal Chemistry**
1977. 80 figures, 9 tables. IV, 157 pages
ISBN 3-540-08366-9

Volume 73
**Organic Chemistry**
1978. 19 figures, 42 tables. IV, 271 pages
ISBN 3-540-08480-0

Volume 74
**Organic Compounds**
Syntheses/Stereochemistry/Reactivity
1978. 37 figures. IV, 133 pages
ISBN 3-540-08633-1

Springer-Verlag
Berlin
Heidelberg
New York

MIX
Papier aus verantwortungsvollen Quellen
Paper from responsible sources
FSC® C105338

If you have any concerns about our products,
you can contact us on
**ProductSafety@springernature.com**

In case Publisher is established outside the EU,
the EU authorized representative is:
**Springer Nature Customer Service Center GmbH**
**Europaplatz 3, 69115 Heidelberg, Germany**

Printed by Libri Plureos GmbH
in Hamburg, Germany